RELIGIOUS EXPERIENCE AND SCIENTIFIC METHOD

By

HENRY NELSON WIEMAN

SOUTHERN ILLINOIS UNIVERSITY PRESS
Carbondale and Edwardsville

FEFFER & SIMONS, INC.
London and Amsterdam

ARCT
URUS
BOOKS ®

PREFACE TO THIS EDITION

This book was my first attempt to solve the problem which has engaged the last forty-five years of my life. This problem can be put in the form of the following question: How can we interpret what operates in human existence to create, sustain, save and transform toward greatest good, so that scientific research and scientific technology can be applied to searching out and providing the conditions—physical, biological, psychological and social—which must be present for its most effective operation?

This operative presence in human existence can be called God in the sense that, with all the many diverse meanings given to the word "God," this creative and saving power has been one of the universal and essential meanings, no matter how diversely the idea of God has otherwise been understood. The word "it" is used not to prejudge the outcome of inquiry by assuming that what we seek cannot be a person but to avoid the opposite presumption that this reality must be a person, no matter what the outcome of the inquiry may be.

The goal of inquiry is not to find what conforms to some chosen traditional idea of God except the one essential element in that tradition, namely, what operates in human existence to create, save and transform toward greatest good.

PREFACE TO THIS EDITION

The greatest good is understood to be the indefinite expansion of the integrated, valuing consciousness of the individual in community with others. In the Christian tradition this has been called the Kingdom of God.

In our time the ever increasingly gigantic power of scientific research and scientific technology has become so great that no religious commitment can save man from self-destruction unless it is of such sort that it can bring the resources of science quite completely into its service. This is not the primary purpose of other prevailing interpreters of religious commitment but it is one essential requirement of religious commitment if we are to be saved from self-destruction. This explains the title of my first book: *Religious Experience and Scientific Method*.

What we seek must be found observably operative in space and time because in no other form can scientific research and scientific technology be brought directly and fully into its service. But this is precisely what must be done if we are to be saved from the misuse of this gigantic power. This first book could only be the beginning of my attempt to solve the problem I have outlined. During the past 45 years I have revised my interpretation of what meets the three requirements: *1*) What truly and observably operates to create the human form of existence beginning with union of male and female cells; *2*) What can expand indefinitely the valuing consciousness of the individual in community with others when our ruling commitment is given to it and other required conditions are present; and *3*)

PREFACE TO THIS EDITION

What does this in such a way that scientific research and technology can be applied to provide these required conditions?

This book, *Religious Experience and Scientific Method*, represents my first attempt to solve this problem but not my last and most mature attempt. In this book I still retained features of the idea of God beyond the reach of empirical inquiry and hence obstructive to the full cooperation of science and religion.

H. N. W.

Grinnell, Iowa
February 22, 1971

PREFACE

The general reader who wants to see the practical results of our position and to avoid the more difficult philosophical discussions should read particularly chapters II, IV, VIII, and IX. These have more of what is called "human interest." Chapters III, VII, X, XI are a little more rigorous. The student and expert, on the other hand, who desire to get the constructive body of the work and test its intellectual accuracy should read particularly chapters I, III, V, VI and XIII.

The chief purpose of this book is to show that religious experience is experience of an object, however undefined, which is as truly external to the individual as is any tree or stone he may experience. It signifies something which extends beyond that space–time occupied by the individual undergoing the experience.

I am very sure that religion must plant itself firmly on the data of sense else it will become the plaything of the sentimentalist and nothing more. If the object of religious devotion is more peculiarly "within" than are the objects of scientific investigation, if it is any more a creature of the human mind than are they, then it will be treated very tenderly, as Santa Claus is treated, being an illusion cherished by children, by the weak spirited and other such who are unable to deal with things as they are. But the strong and intellectually alert will have nothing to do with it. Religion will continue to lose, as it has already lost, intellectual standing. And just as far as it loses intellectual standing it will be given over to sentimental gush and serve its chief purpose in providing a means for self deception to those who want to play the ostrich.

It would be presumptuous to call this work a philosophy of religion, but it is a first step in that direction. There is a reciprocal relation between science and religion which I try to trace in parts one and two, intending to show that neither can maintain itself in adequate manner without the other. In part three I endeavor to clear away from the face of religion what I believe to be certain present-day misinterpretations of it and to state my own view of its function in human living.

Among teachers to whom I owe much, first mention must be made of Professor Wm. Ernest Hocking. How far I may have departed from his teaching I do not know, but I do know that I have taken heavily from him. For introduction to the spirit of scientific method, and enthusiastic appreciation of it, I owe most to Professor Ralph Barton Perry. The influence of the writings of Professor John Dewey will be apparent throughout the following pages. For most recent encouragement in bringing the present work to publication I want to thank Dean Shailer Mathews, Professor John A. MacIntosh and Reverend Hugh Kerr. Of course none of the persons mentioned can be held responsible for the views here presented.

While for the most part the material here published is new, portions of it have appeared in the *Journal of Religion,* the *Journal of Philosophy* and the *International Journal of Ethics.* Acknowledgment is here made to the editors of these journals for the use of this material.

In chapter VIII I have made quite extensive quotations from the *World Tomorrow* and the *Atlantic Monthly.* I want to thank these journals for the permission they have given to borrow from them; and especially to thank Mr. Paul Green for the use of his story "The Devil's Instrument" published in the *Atlantic.* Chapter XII appeared in the *Journal of Religion,* Vol. 5, Sept., 1925, under the heading "Religion in Dewey's Experience and Nature."

H. N. W.

INTRODUCTION

Whatever else the word God may mean, it is a term used to designate that Something upon which human life is most dependent for its security, welfare and increasing abundance. That there is such a Something cannot be doubted. The mere fact that human life happens, and continues to happen, proves that this Something, however unknown, does certainly exist.

Of course one can say that there are innumerable conditions which converge to sustain human life, and that is doubtless the fact. But in that case either one of two things is true. Either the universe is a single individual organic unity, in which case it is the whole indivisible universe that has brought forth and now sustains human life; or else certain of these sustaining conditions are more critically, ultimately and constantly important for human welfare than are others. According to the first view God would be, or involve, the whole universe; according to the second he would be those most important conditions which, taken collectively, constitute the Something which must have supreme value for all human living. The word God, taken with its very minimum meaning, is the name for this Something of supreme value. God may be much more than this, but he is certainly this by definition. In this sense, with this minimum

9

meaning, God cannot be denied. His existence is absolutely certain. He is simply that which is supremely significant in all the universe for human living, however known or unknown he may be.

Of course this statement concerning God proves nothing about his character, except that he is the most beneficent object in the universe for human beings. He is certainly the object of supreme value. Nothing is implied by this definition concerning personality in God; but neither is personality denied. In fact, personality is by no means a clear and simple term. But two things are made certain: his existence and the supremacy of his value over all others, if we measure value in terms of human need.

But such abstract reasoning concerning God accomplishes little save to clear the ground, remove misunderstanding and prejudice and open the way for the more vital problem.

We believe nothing is more important at the present stage of human thought than to define God in terms of concrete experience. Failure to do this has led some of the finest scientific thinkers of our time to regard religion as superstition and nothing else. E. Rignano is typical of such thinkers and such an attitude. After tracing the many attempts that have been made to define and demonstrate God and the immortality of the soul by means of "metaphysical reasoning," and finding all these attempts futile, being nothing but "defense reactions" by which men try to make themselves think the world to be as they would like to have it, and concealing the nonsense in all their intellectual constructions by means of the vagueness and obscurity of their thought, he goes on to say:

> These mystical minds—which will always exist —will always continue to try to realize and systematize their aspirations and their dreams in transcendent constructions, ever new, ever different, ever vain; pale reflections of the great systems of the past, the last gleams of a great human illusion that has vanished. And these metaphysical speculations, old and new, will in their entirety constitute a great epic handing down to posterity the exploits of the tragic revolt, worthy of Prometheus, which the infinitely small microcosm has dared and will dare again to attempt against the infinitely great macrocosm.[1]

He contrasts with this "metaphysical," theological and "mystical" reasoning the intellectual work of the positivist and scientist who goes to experience for the truth and whose thinking consists altogether of experimentation; who discovers the truth partly by actually carrying out experiments on nature by physical manipulation, but even more by combining in imagination innumerable physical experiments that can only be combined in such great numbers by imagination and thus be made to reveal new facts otherwise never to be discovered. This classification, multiplication, and unique combination of experimental operations in thought by means of concepts is the greatest achievement of the intellect, the only means by which truth can be brought to light, and the only

---

[1] *The Psychology of Reasoning*, p. 261.

way in which human aspirations can be rendered effective in this world or wrought into the body of existence.

In all these claims which Rignano makes for scientific thought, we believe he is correct. When he points out the error of those who turn away from experience to find God by metaphysical speculation, we believe he is rendering religion a service. Metaphysics in the sense of that reasoning which adjures experience and the conclusions of scientific thought, is futile, as Rignano says.

Furthermore, we agree with him in his condemnation of mysticism, if mysticism is used in the sense in which he uses it. Perhaps his use of the word is the only legitimate use; but we believe our usage is justified. However, if it is not, our thought must be tested by its actual content and not by the fact that we have given an unconventional meaning to the word.

To clarify our usage of mysticism and distinguish it from other meanings attached to it, let us consider how Rignano applies the word. He uses it in two slightly different senses.

In the first place he means by mysticism the reverence or emotional glow which may attach itself to certain words or phrases when these have lost all intellectual significance. When a word or other symbol has ceased to signify any fact of experience whatsoever, and precisely because it does not signify anything which will contradict or render foolish and unwarranted the enthusiasm which attaches to it, as a mere word without intelligible meaning, it stirs the mystic to emotional raptures. In other words, by mysticism Rignano means sentimentality. "Certain terms thus become in time pure sounds, no longer evoking intellectual representations, but only emotions; and not certain particular emotions relating to a well-determined object, but 'general emotions,' similar to those aroused by a 'series of musical notes in the minor mode.' "[2]

Now there is no question but that this kind of "mysticism" does occur. There are people who maintain such an attitude toward certain intellectually meaningless symbols, and the value of the symbol can be preserved for them only by keeping it intellectually meaningless. If one wishes to call this mysticism, of course he can. It is a legitimate use of the word. But it is not the sense in which we shall use the word.

There is a second sense in which Rignano uses mysticism, closely allied to the first. It is the belief that back of the world of sense there is some "noumenal reality" inaccessible to experience, but nevertheless existent. ". . . . any act of mysticism does just consist in admitting the existence of something mysterious which is not capable either of coming under the observation of any of our senses or of being imagined by means of sensible elements combined together in any fashion."[3]

It is interesting to note that he accuses mathematicians, many and prominent, of being mystic in this sense. Such mysticism rises among mathema-

[2] *loc. cit.* p. 256.
[3] *loc. cit.* p. 185.

ticians when they begin to use mathematical symbols which refer to such infinities of complex operations that they lose the connection between these symbols and the empirical operations signified. Consequently, when they operate with the symbols to produce results which are mathematically correct and yet seem to have no existence in the world of experience because they have lost the reference these symbols have to empirical operations, they think they are dealing with some transcendental reality. To demonstrate the error of this view Rignano very acutely traces the higher branches of mathematics showing how they refer, directly or indirectly, but always ultimately, to the world of experience.[4]

Here again we have a kind of "mysticism" that does certainly occur. There are many people who feel that they are leaping into a transcendental sphere beyond the reach of all experience when they operate with symbols which, so far as they can see, do not refer to anything in experience. We agree with Rignano in claiming that when we deal with such symbols either one of two things is true. Either the symbols do refer to facts of experience which we fail to discern, or else they refer to nothing at all and hence are meaningless. In neither case do they introduce us to the transcendental. Of course a symbol may refer to experience in a very indirect fashion, by referring to other symbols and these to others and so on. But if the ultimate object

---

[4] Our own Cassius Keyser would doubtless come under the head of such mystics. See his *Human Worth of Rigorous Thinking*. Also Spaulding with his *New Rationalism*.

of reference be not some phase of experience, then this mutual reference of symbols to one another merely serves (if it serves any good at all) to define and distinguish the symbols from one another, so that they can be used more effectively in designating some object of experience. But the only objects we can know are the objects of experience.

We trust that our use of the word mysticism will not be confused with either of these meanings which have so frequently been attached to it. By mysticism we mean a certain way of experiencing the world of empirical fact, and nothing more. It is the sort of experience described by William James, for instance, in his essay on "A Certain Blindness in Human Beings," quotations from which will be found in chapter III following.

There are times when men, with a partial suspension of thought processes, become aquiver with the vast fullness of sensuous experience that rains down upon them. This is the mystic state. It may be brought on by symbols and in many other different ways. But it is not a thinking state; it is merely a form of immediate experience. It cannot yield knowledge until it is correctly interpreted. Its true meaning must be brought to light by intellectual operations which are not mystical. In the mystic state one does not think, he does not cognize, he is simply immediately aware—of what? Of the fullness of some concrete experience. Since mysticism is not a thinking state, the definition and description of the religious object, God, cannot be the work of mysticism, although mysticism may supply the datum through which intellect may discover God.

NOTE: The reader who wishes to omit the more theoretical portions and read only the parts of general interest should pass over chapter I and read chapters II, IV, VIII and IX.

CONTENTS

17

ERRATA

PAGE 373, LINE 13: *For* and *read* und
PAGE 374, LINE   2: *For* and *read* und
PAGE 374, LINE   4: *For* and als Naturmmthologie
                    read und als Naturmythologie
PAGE 374, LINE   9: *For* eigeneen *read* eigenen
PAGE 374, LINE 11: *For* and *read* und

PART I
WHY RELIGION NEEDS SCIENCE

Religious Experience and Scientific Method

CHAPTER I

SCIENCE AND OUR KNOWLEDGE OF GOD

Two views have been held concerning the way we know God. One has asserted that we must know God just as we know any other object; that there are no other powers or faculties of knowledge except those by which we know ordinary objects; and that we must know God as we know trees and houses and men or else not know Him at all. The other view has tried to show that knowledge of God is a special kind of knowledge; that there is a certain feeling, inner sense, eye of the soul, instinct, or intuition, faith, spiritual organ, moral will, or what not, which has God as its special object; that trees, houses and men may be known through interpretation of the data of sense that God is discerned in this special and peculiar manner.

Now there are two senses in which one may refer to feeling, intuition, faith, moral will, etc., as means to knowing God. By these terms one may mean merely to designate certain distinctive kinds of experience which provide the data that may lead to the knowledge of God if correctly interpreted. If that is all that is meant, then the assertion amounts to the first of the two above mentioned views. If that is what is meant, we know God as we know any other object, for that is the manner of all knowledge. But if faith, feeling, intuition, and moral will are represented as giving us an immediate kind of knowledge in which there is no need for the analysis and interpretation of immediate experience by means of concepts, then the assertion amounts to the second of the two above mentioned views. Then the claim is that a certain kind of immediate experience gives us knowledge without the intervention of any further intellectual processes.

This second view we hold to be false. We believe it erroneous, in the first place, because it identifies knowledge with immediate experience. Immediate experience never yields knowledge, although it is one indispensable ingredient in knowledge inasmuch as it provides the data from which knowledge may be derived. We hold this view wrong, furthermore, because it resorts to a peculiar and mysterious faculty, as though every special kind of object must have a special kind of faculty for discerning it. These mysterious faculties of discernment have long since been regarded as mythical by psychology and epistemology so far as all ordinary cognition is concerned. To cling still to such a view with respect to discernment of God is to put the knowledge of God outside the field of scientific knowledge, where it can be neither examined or tested. Such a position is fatal to religion.   It means that knowledge of God will be calmly ignored by all those who are interested in scientific thought.   As a matter of fact, we believe it is precisely because this view has prevailed that knowledge of God has been so widely ignored in scientific circles.   And by scientific circles we mean scientifically inclined philosophers, and all who are dominated by the scientific method, whether they be professional scientists or not.

All knowledge must depend ultimately upon science, for science is nothing else than the refined process of knowing.   Scientific method is simply the method of knowing.   We call it scientific only because it has been deliberately developed for the purpose of guarding against error.   All knowledge is scientific except in so far as it has not developed a method for discriminating accurately between the false and the true.   Ordinary knowledge is distinguished from the scientific only because of its vagueness and its undetected fancies and illusions. The knowledge of God must be ultimately subjected to scientific method.   We say ultimately rather than immediately because, as we shall see later, science has not yet developed a method adequate to deal with the more complex data of experience.   Physics became a science long before psychology because its data were so much more simple.   Sociology has scarcely yet attained the status of a science because its data are so complex. The datum of religious experience is so exceedingly complex that no method has yet been devised which is fit to treat it scientifically.   But we are working

in that direction. In the meantime men can have
acquaintance with God without accurate knowledge of Him, just as men could have acquaintance
with matter long before there was any science of
physics to give them scientific knowledge of it; just
as men had acquaintance with food long before
there was any scientific knowledge of the nature of
food. Our knowledge of matter and life and mind
is more or less scientific. Our knowledge of society
is becoming scientific. We do not yet have any
knowledge of God that can be called scientific.
But for centuries our knowledge of the object of
religious experience has been growing more scientific.

But before we can consider adequately the relation between science and our knowledge of God we
must get as clear an idea as possible of the nature
of cognition.

Immediate experience, we have said, does not
necessarily yield knowledge at all; much less does
it necessarily yield true knowledge. Our hand may
brush a table in the dark and yet we do not know
it. We may not interpret the experience at all.
We may not know that our hand has touched anything, our mind being turned to other things. We
have had a genuine experience of the table, an experience, however, in which there is no cognition
or knowledge. The same, of course, applies to visual sense data or any other sensation or combination of sensations. The image of the table may fall
upon the retina of my eye and I be unaware of it
or interpret it wrongly, thinking it to be a shadow.

While immediate experience is not identical with knowledge and does not necessarily yield knowledge, yet our knowledge of the concrete external world, including other minds, is derived from immediate experience. We know an object when we are able to designate certain sense qualities having a certain order in time and space. When we experience one or more of these sense qualities in certain temporal and spatial relations to other sense qualities, we are able to infer that the object before us is of a certain sort, a table or chair or what not. We test this inference by exposing ourselves to further sense qualities. If these further sense qualities are of the sort that properly pertain to the inferred object occurring in that order in space and time which is proper to the object in question, we know that our inference was correct. Our knowledge is then fairly certain. All the elaborate tests of scientific investigation depend ultimately upon this corroboration of inference by means of sense data. Of course the situation may be so familiar that we infer instantly from a bit of given experience what the object is, and do so with a high degree of certainty.

We want to make a distinction between knowledge by acquaintance and knowledge by description, but not using these terms in exactly the same sense as William James, who coined the phrases. Knowledge by acquaintance, according to our usage, is of that which has been experienced by some one, or presumptively could be if the right kind of organism could be placed in the right situation. Knowledge by description, on the other hand, is of that which could not be experienced by any organism or mind whatsoever, because it does not refer to any of the data of experience. Mathematical points and lines are such objects of knowledge. They do not belong to the realm of experience at all, although they may be related to experience by means of what Whitehead calls "extensive abstraction." All geometrical, mathematical and purely logical entities are outside the realm of experience. Our knowledge of them consists of concepts which refer to other concepts and not to any data of experience. All number is of this sort, number being no object or group of objects which can be experienced, but being a "class of classes," according to Bertrand Russell.

Now in both these two kinds of knowledge, that of acquaintance and that of description, knowledge requires a whole system of concepts. An isolated proposition does not yield knowledge any more than an isolated datum of experience. But the difference between the two is that in description the single concept refers to a system of concepts and nothing more; while in knowledge by acquaintance the system of concepts which are thus brought into play serve to designate certain data of experience. In accurate scientific knowledge this system of concepts serves to define the order in space-time in which certain data of experience must occur in order to constitute a certain object. Scientific knowledge consists in designating this order of experience plus, generally, certain hypothetical entities which are required to fill out the fragmentary experience. But in purely descriptive knowledge there are no such data of experience. In purely descriptive knowledge there are no hypothetical entities to fill out the ragged edges of experience, because no experience enters into the object of such knowledge. Such knowledge is simply the interlocking of a perfectly consistent system of concepts without regard to any experience whatsoever. Any perfectly consistent set of propositions which have been built up without regard to any particular experience, would be knowledge by description. Some would say that such consistency could not yield knowledge of anything save of the necessary logical requirements of thinking.

Now this distinction between these two kinds of knowledge has a very direct bearing upon knowledge of God. It raises one of the most important questions that bear upon this matter. Is our knowledge of God knowledge by acquaintance, or is it purely descriptive? Is God an object that enters into our immediate awareness, or is He only an object of speculation, known only through the logical consistency of propositions, which must be the form of all accurate knowledge, but known through a logical consistency which does not define any object entering our immediate awareness? Is he an object of possible experience, or is He purely a system of concepts? A great deal of religious thinking has interpreted God as a system of concepts, and that only. They have not necessarily denied that God was an object of possible immediate experience, although Kant did just that, because he was clear-headed enough to see that that was the inevitable outcome of his position. But many, not so clear in thought as he, have sometimes seemed to imply that a system of concepts could be experienced the same as a horse or a cow. But concepts cannot be experienced in the same way as fire or earth. Can God be so experienced? Either God is an object of sensuous experience, or else He is purely a system of concepts and nothing more. All attempts to escape this dilemma must result in confusion and befuddlement, if not in actual superstition.

We believe that advance toward scientific knowledge of God has been delayed by failure to recognize this sharp distinction and insist that God must be either the object of sensuous experience, or else a system of concepts and nothing more. In any case, God must be known through a system of concepts, for there is no other way of knowing. But if He be an object of acquaintance, the system of concepts refers to certain experiences, while if He be not, the concepts only refer to one another. If He be not an object of sense experience, He cannot be scientifically known. Knowledge of Him then becomes purely a matter of logic, a matter of reducing certain concepts to perfect consistency, but without any attempt to use them to designate any data or datum of experience. Logic has its part to play in all knowledge, of course; in knowledge by acquaintance as well as in knowledge by description. But knowledge by acquaintance is logic plus, while knowledge by pure description is not. In knowledge by acquaintance logic merely assists by devising those concepts which can best serve to define the character and significance of the data of experience.   In knowledge by pure description logic does everything.   It is significant to note that Whitehead and Bertrand Russell have discovered that the ultimates of logic and mathematics are identical.

God, we claim, is not a logico-mathematical entity. He is an object of immediate experience. Mystical experience of the sort we shall portray, akin to aesthetic and social experience, but clearly different from both, must be scientifically interpreted if we are to know what God is. If by God we mean the object of such experience, without any further attempt to describe his character, then there can be not the slightest doubt in the world that God exists. For there can be no question about the reality of religious experience; and all experience is the experience of something. Religious experience is just as real as any experience; just as real, for instance, as the experience of a human beloved, or color, or sound, or the experience given in dreams and hallucinations, or those experiences which are said to be of trees and stones. All experiences are of equal reality. The only question that can be raised anent them is concerning their precise character and their significance. All experiences signify something. All experiences are experiences of some object or other. The only question is: What object? We often do not know the nature of the object we experience. In hallucinations we plainly are mistaken concerning the object experienced. But that does not belie the fact of some object. It could not be an hallucination if there were no true object to give the lie to the hallucination. Absolute scepticism concerning existence of the objects of experience is impossible, even though one must invent an "animal faith" to escape from it.[1]

Why are people so uncertain concerning their concept of God if He be an object of experience as genuinely as tree and hill and stone? One reason we have already indicated. The datum of experience is so complex that no scientific method has yet been devised which is able to deal with it. But a more immediate reason, a corollary to this, is that religious experience has not been adequately distinguished from many other forms and phases of experience. In our experience of God there is a merging of many experiences, and just that form which gives us the datum signifying God is not clearly distinguished from that which gives us our knowledge of earth and sky, and fellow man and social group. Now of course this merging of many phases of experience is inevitable, not only in the case of God, but in the case of everything else. When we have a visual experience of a stone we also, in the same situation, experience the light and hence the sun, also the earth and sky, and heaven knows what all. All these are merged in our experience of the stone. But in the case of the stone we have learned how to distinguish between that phase of experience which can be said to be data

---

[1] See Santayana, *Scepticism and Animal Faith.*

pertaining to stone, and that phase which is data pertaining to sun, etc. There is probably an infinite wealth of data merged in any ordinary experience, but we have learned to distinguish and select from this infinite wealth those data which enable us to adapt ourselves to stones and trees and hills and make inference concerning them. But with respect to our experience of God we have not learned to do this with so much clarity.

Religious experience is one of the most ancient and widespread of all experiences, but the distinctive merging of data constituting the total datum of religious experience has never been satisfactorily defined. The same is true of that datum of experience from which we derive our knowledge of other minds. Perhaps religious experience and experience of other minds belong to the same category. Because the data pertaining to God and other minds is so much more complex than that pertaining to stones and trees, it is very natural that we should be much slower in clarifying the former than the latter. Hence sociology, and above all, scientific theology, will be among the latest of all sciences when they finally arrive. But until they do arrive, our social intercourse with God and other minds will be full of delusions, blunders, vague fancies, and blindness to fact. Until these objects of our immediate experience become more intelligible, our society and our religion will be full of error and confusion, as they certainly are to-day. Since science has clarified the data out of which we develop our knowledge of physical nature, our knowledge in this field of

experience has proceeded by leaps and bounds, and
our adjustments to the processes of nature is one
of the marvels of the age. But in pitiful contrast
stands our knowledge of those objects which are
experienced in human society and mystic worship
—fellowman and God. How confused and how
full of illusions and blunderings are our social and
religious adjustments as compared to our physical,
chemical and biological. The difference is that
science has entered the field of physics, chemistry
and biology. It has only begun to enter the field of
society; and the field of religion it only approaches
from afar off.

Some might claim that it is as absurd to speak
of science entering the field of religion as to speak
of it entering the field of alchemy or astrology. But
that is precisely the point. Science did enter the field
of alchemy and astrology and there resulted the
sciences of chemistry and astronomy. There was a
time, no doubt, when mathematicians ridiculed the
notion that the blundering efforts of the alchemists
could ever develop into a science. But they did.

There can be no question about the fact of religious experience, as we have already indicated. It
is just as much a fact as any other kind of experience—as eating or dreaming or loving, for instance.
The only questions are: What sort of object is experienced? And second: What is the significance
for human living in such experience? These questions can be answered satisfactorily only by science,
if by science we understand merely that method by
which truth and error are discriminated and knowl-

edge verified. Since that is the sense in which we are using the word science, it becomes a truism to say that these questions can be answered by science.

In moving toward a more adequate, *i.e.* a more scientific, knowledge of God, even though we approach from afar off, three things are required: (1) a clarification of that type of experience which can be called distinctively religious; (2) an analysis or elucidation of that datum in this experience which signifies the object being experienced (God); and (3) inference concerning the nature of this object.

It is true there are scientists, and others, who ridicule any attempt to reach verifiable knowledge concerning the existence and nature of God. That seems to be the attitude, for instance, of C. D. Broad in his contrast between "critical" and "speculative" philosophy.[2] But we must remember that this is a very old old story. Every new science has been ridiculed by those whose bent of thought has been shaped by some older and better established science. Chemists were once ridiculed when they claimed to be able to reach any knowledge of facts by other data than those recognized by physics. Biologists even yesterday, and perhaps in a few quarters even today, were ridiculed when they claimed to deal with any other data than those which physics and chemistry can treat. Psychologists are ridiculed when they claim to deal with any data other than the biological, physical and chemical. While all the sciences, forgetting the precarious days of their own

2 Broad, C. D., *Scientific Thought,* chap. I.

origin, turn with scorn upon sociology. With these facts in mind it ought not to disturb us when the quest of God as an object of knowledge is held up to contempt.

We have said that the first step in moving toward an adequate or scientific knowledge of God is to define the character of distinctively religious experience. We shall do this more fully in later chapters. But all such definitions must be more or less tentative, for the science which has God as its object has not yet arrived. But much work has been done and is being done in this field. William James in his *Varieties of Religious Experience,* has taken us a long stride in the direction of distinguishing religious experience from other forms of experience. William Ernest Hocking has carried us on farther still in this direction with his *Meaning of God in Human Experience.* Many others might be mentioned. If many individuals contribute to the work of distinguishing, recording and interpreting religious experience, and if the great religious experiences of history are preserved with some degree of accuracy, it may be that human history will culminate in an adequate idea of God. We can imagine no loftier culmination to the life of the race.

The most common human appeal to God is in the hour of bewilderment, when the individual (and often the group) feels baffled and defeated. It is when he is not sure of himself that he turns to God; when he is in doubt and yet feels the urgency of action; when he does not know which way to turn and yet feels that he must turn some way.

Above all it is when one has staked all his life's success and happiness upon some enterprise and feels it threatened with disaster or actually ruined. It is at such a time that one has a sense of God. Here, for instance, is an example recorded by Rufus Jones.

> I had a friend who went alone one day to consult a famous London doctor. My friend was very highly gifted and was at the time just beginning to reveal unusual literary powers and he was at the opening stage of a promising career in business. At the same time he was coming to be recognized as the spiritual leader of the younger section of his religious fellowship. Everything which makes life rich and great was before him. The doctor gravely and with almost killing frankness told him that he was the victim of a subtle and baffling disease which would destroy his hearing and his sight and would eventually seriously affect his memory. He came down the stairs of the doctor's office and stood almost stunned on the curb of the street, realizing that all the large plans for his life had collapsed like a child's house of blocks. Suddenly as he stood there, waiting to decide which way to go, he felt as though he was enveloped by the invading love of God and filled with a sense of unutterable peace. There emerged within him a source of energy sufficient to turn his primary tendency to despair into steady consciousness of hope and joy which lasted throughout his life and gave him extraordinary power and influence.[3]

[3] Jones, Rufus M., *Fundamental Ends of Life*, p. 106.

The Freudians, of course, would say that this was the sudden emergence from the subconscious, of a phantasy; it was a "flight from reality" carried out by the mechanisms of the subconscious. This Freudian view and the general standpoint of Freudian psychology we shall treat in Part III. But, for the present, let us take another instance, not so readily subject to this widespread Freudian theory. Geo. A. Coe has given us the following:

> The Titanic survivors who were rescued by the Carpathia, so Stanton Coit, an eyewitness, relates, seemed not to be stunned and crushed but "lifted into an atmosphere of vision where self-centered suffering merges into some mystic meaning. . . . We were all one, not only with one another but with the cosmic being that for the time had seemed so cruel." Still more significant is Professor James' analysis of his own attitudes and those of others on the occasion of the great California earthquake, which overtook him at Stanford University. "As soon as I could think," he says, "I discerned retrospectively certain peculiar ways in which my consciousness had taken in the phenomenon. These ways were quite spontaneous and, so to speak, inevitable and irresistible. First, I personified the earthquake as a permanent individual entity . . . . Animus and intent were never more present in any human action, nor did any human activity ever more definitely point back to a living agent as its source and origin. All whom I consulted on the point agreed as to this feature in their experience. 'It expressed intention,' 'It was vicious,' 'It was bent on destruc-

tion,' 'It wanted to show its power,' or what not. To me it simply wanted to manifest the full meaning of its name. But what was this 'It'? To some apparently, a vague, demoniac power; to me an individualized thing."[4]

What precisely is it, in psychological behavioristic terms, that occurs at such a time? In bewilderment the established system of habits are frustrated. There is a quickening of a great many different impulses, new and old, a flinging out aimlessly of innumerable unorganized responses which, for the most part, do not reach the stage of overt action. We must imagine a case of extreme bewilderment, not merely where one considers a few rather well defined alternatives. An illustration would be a business crash where one's business career is apparently ruined, or disappointment in love, or an accident which causes one to give up his chosen calling in life, or the death of some beloved who was the center of one's life. At such times one's attention ceases to be focused on certain definite objects and becomes diffusive. The ordinary objects of response being taken away, and at the same time the whole organism pervasively stimulated, brings into play innumerable impulses without any determining adjustment or established pattern. It is a state where one must necessarily be aware of concrete, unanalyzed masses of experience that surge in upon one. We do not mean that there is any conceptualized cognition of these masses of experience. That is exactly what we do not mean, for we have all

4 Coe, Geo. A., *Psychology of Religion*, p. 100.

along insisted on the distinction between awareness of immediate experience and clear knowledge of an object.

Among the lower animals, when the established course of conduct is blocked and there is no pattern of hehavior to guide, there is also an aimless flinging about. But among lower animals this flinging about always consists in running through a certain repertoire of more or less definite and established patterns of hehavior. Sometimes this is all the human does. But in the case of the human the great number of different impulses, some of them perhaps never before aroused, that may be quickened in such an hour, may produce that undefined awareness of the total passage of nature, the undiscriminated event. The bounds of awareness are greatly widened. The ordinary narrow and routinized selectiveness of attention is broken down, and instead of attending only to a few familiar data to the exclusion of all else, one becomes aware of a far larger portion of that totality of immediate experience which constantly flows over one. This is precisely what we have previously defined as the unique datum of religious experience.

Now anything which breaks up the established system of response by which we react to the habitually selected data, and throws our responses into confusion, may produce in us that simultaneity of innumerable responses by which we become aware of this movement of total experience. Whenever this befalls us we have that which at least those who have had appropriate religious training recognize as the experience of God. Obstruction and bewilderment, we have said, is one thing which brings this on. But of course it is not the only thing. Gazing at the Grand Canon of the Colorado River may break down in one all his habitual systems of response and produce that simultaneity of innumerable impulses by which one becomes more or less aware of the unanalyzed and unsifted stream. Falling in love may produce this experience, if it breaks down one's established system of response and fuses into one, total response innumerable impulses, new and old, through which one becomes aware of the flow of experience. Something of the same sort may be brought about by profound and affectionate interchange of thought between intimate friends through which the deeps of response are stirred; or by the effect of an excited crowd upon the individual—a revival meeting, for instance.

Perhaps the most common experience of God, however, has nothing to do with such great crises as we have indicated. This state of diffusive awareness, where habitual systems of response are resolved into an undirected, unselective aliveness of the total organism to the total event then ensuing, comes softly in the quiet hour, unannounced and unrecorded. Such experiences, we believe, are very common to many people, but here is the confession of H. G. Wells.

> At times in the silence of the night and in rare lonely moments, I come upon a sort of communion of myself and something great that is not myself. It is perhaps poverty of mind and language which

obliges me to say that this universal scheme takes
on the effect of a sympathetic Person—and my
communion a quality of fearless worship. These
moments happen, and they are the supreme fact
of my religious life to me; they are the crown of
my religious experience.[5]

This indefiniteness, inadequacy and uncertainty
of the concept of the object experienced, which
Wells indicates, shows what we have all along
insisted on. Religious experience gives us an indubitable datum, which is more certain than knowledge, but it does not give us knowledge of God,
except as it is correctly and exhaustively interpreted.
One might have immediate experience of a bell and
not know anything about a bell, have no concept
of a bell, and never know that he was experiencing
a bell. So also with the experience of God.

With respect to the knowledge of God we feel
that it is very important to distinguish between
knowledge that we have experienced God and
knowledge of what sort of object God is. Of these
two stages of knowledge, the knowledge that we
have experienced God is primary and most important. The value of clearly distinguishing the datum
of religious experience is that it enables us to know
that we experience God and when we experience
God. All the values of religion *per se*, as distinguished from theology, religious philosophy and
religious ethics, is to be gotten from this experience
of God. The values of this religious experience can
be immeasurably enhanced by proper ideas about

---

[5] Wells, H. G., *First and Last Things*, quoted by Rufus Jones, *op. cit.*

God, hence none can prize more highly than we a good theology, philosophy and ethics. But all these ideas about God cannot enhance the value of the experience if we do not have the experience. Without the experience we have no religion at all. With the experience we may have a very bad sort of religion because of our false or inadequate ideas about God: namely, our philosophy, ethics and theology. But no matter how excellent these latter may be, we have no religion at all if we do not have the experience. We must develop a better theology, philosophy and ethics. But most important of all we must cultivate religious experience, which is acquaintance with God.

It is neglect of religious experience as such which is our chief danger in this age of scientific method. The more rigorous the scientific method, the more need have we of religious experience and the more need has science of this experience to keep it ever youthful and growing. But when the unique character of religious experience is not clearly discerned and its indispensable value recognized, there is danger that religious experience will be ignored and neglected in the wave of enthusiasm for scientific method which is now rising among us.

When the value of religious experience is not put on an equal basis with that of scientific method and upheld as an indispensable rejuvenator of science, there is no danger that men will cease to have religious experiences or cease to seek such experience. The need of religion is too deeply seated in our nature for that. But the danger is that the interpretation of this experience, the philosophy, theology and ethics, which give character and direction to the vision and energy derived from religious experience, will degenerate. The danger is not that we shall cease to be religious, but that the quality of our religion will decline. Men will not cease to experience God, but their understanding of God will become more and more inadequate to the requirements of our life. Religion needs science as much as science needs religion. The evil effects of the divorce between the two is shown increasingly by the innumerable types of religion that are developing among us. Note the esoteric literature, mental healing, star reading, alchemy, the cabala, transcendental magic, the higher and lower mysticism, Eastern scriptures, Spiritualism, Seventh-Day Adventists, the Nazarenes, Pentecostals, Church of God, Swedenborgians, Pillars of Fire, Theosophists, Buddhists, New Thought, of many stripes, as Divine Science, House of Blessing, Home of Truth, Rawson Teaching, Circle of Divine Ministry, Practical Christianity, Metaphysical Library, Unity Center, Practical Psychology, Fundamentalists, Modernists, etc., etc. We do not mean to lump all these together as equally bad or equally good. We only mean to say that such an amazing confusion of religions shows that there is something wrong in our treatment and interpretation of religious experience. We do not deny but that members in all these sects may have genuine experience of God. But they cannot all interpret and apply the fruits of that experience equally well. This confusion in the interpretation and application of the experience will go from bad to worse if science does not recognize the nature and value of religious experience and affiliate itself more closely therewith.

What we have said of religious experience smacks strongly of Bergson. But we would be greatly misunderstood if our thought were identified with his. In order to avoid this confusion we must show wherein we differ from him. There are three points of difference.

In the first place we do not agree with Bergson in saying that instinct gives us an awareness of the unanalyzed and unselected mass of experience. Instinct is simply the operation of certain automatic mechanisms of behavior. These mechanisms do, of course, determine the objects of our attention. But they are just the opposite of what Bergson says they are in this respect. They are highly selective. They do not render us responsive to, sensitive to, or conscious of, the unanalyzed flow of experience. On the contrary they are the first steps which the organism takes in selecting from the mass of stimuli which assails it, those particular elements which are of practical importance. If intellect deals with experience in the interests of practice, certainly instinct does so no less.

Our second point of difference has to do with identifying what Bergson calls intuition with knowledge. We are never fully aware of the unanalyzed and unselected mass of experience in its original continuous flow. But we believe Bergson is right in saying that we may have various de-

grees of awareness of it. We have tried to show some
of the situations which give rise to this awareness.
But where Bergson makes his mistake, we believe, is
in identifying this immediate awareness of experience with knowledge, and treating it as a peculiar
kind of knowledge different from intellectual cognition and designated by the term intuition. This
only leads to confusion. Our awareness of the continuous flow of experience is not different in kind
from our immediate awareness of a touch or a
sound, except that in the former the work of analysis and selection of data has not been carried so
far as in the latter.

Our third point of difference has to do with.
metaphysics. Because of his initial error of confusing immediate experience with knowledge, Bergson
is led to the conclusion that through our awareness
of the continuous flow of experience we have in-
tuitive knowledge of ultimate reality and that this
ultimate reality is a continuous stream of experience
without thought and without purpose, but which is
ever evolving into something further, the something
further being wholly undetermined and unknown
until actually achieved. In contradistinction to this
we have maintained that our immediate experience
of God is merely a datum, and taken by itself alone
gives us no knowledge concerning the character of
God. Because this datum is a continuous, unseg-
mented flow of experience, we cannot immediately
jump to the conclusion that God is a universal, unthinking, unpurposing flow of experience, any more
than we can conclude that a chair is a disembodied

pressure because our immediate experience of chair is that of pressure.

The experience of God by itself alone does not constitute religion. One must interpret that experience before he has a religion. And from time immemorial man has given some interpretation to his experience of God. This interpretation has been just as crude, grotesque and diverse as his interpretation of other experiences. He is subject to the same errors in the interpretation of this datum as in any other attempts to understand that which befalls him. He can hope to correct these errors only as he subjects his interpretation to the only methods we have for detecting error and revealing truth.

While we have no science of God in the strict sense, yet science has not been without its influence upon our ideas of Him. Advancement in one branch of knowledge cannot occur without influencing all other branches, and generally in the way of correcting error. So our ideas of God are undergoing, and have for centuries undergone, correction under the influence of science and pre-scientific investigation and thought.

Finally it must be made clear that any science or near science of God which may ever be developed will have all the limitations of scientific method, such as we shall later note. Scientific method as applied to God, or to anything else, tends constantly to refine its data, excluding everything irrelevant and constraining imagination to follow certain prescribed forms. But what can be said to be irrelevant to God? Perhaps nothing is irrelevant to Him. But the nature of scientific method requires simplification of data and therefore necessarily the casting out of much that is irrelevant. Hence the science of God will need to be counterbalanced by the mystic experience just as every science requires such correction and rejuvenation. We are already very familiar with the growing sterility of theology unless renewed and corrected by such immediate and mystic experiences. Nevertheless, we need a science of God as we need a science of any other important object with which humans must deal. Such a science will have all the dangers of sterility and vital impoverishment which we find in all sciences; and it will be peculiarly dangerous because it will lay its bonds upon the ultimate sources of that mystic experience which is the greatest counter balance to all scientific method. But we trust that as science draws closer and closer to our knowledge of God we shall learn how to correct its abstractions by means of concrete experience. And above all, as we shall try to show in a later chapter, love solves the problem of how to combine the truth of science with rich concrete awareness of the object known.

But the fact that we have no science of God and possibly may never have must not be taken to mean that we have no true knowledge of Him. Without a science we have no accurate method of verifying our ideas and certainly distinguishing between truth and error. Therefore it behooves every one to hold his ideas concerning God with an open mind, being absolutely certain only of the fact that God is, on the grounds already mentioned. But the fact that we cannot verify our ideas about God, nor distinguish with certainty between truth and error, in no way implies that we do not possess much truth about Him.   Men always have much knowledge more or less true, long before it can be scientifically verified.   The same is true of our knowledge of God.

CHAPTER II

SCIENCE THE CORRECTIVE OF RELIGION

Christianity needs science to free it of sentimentality. Sentimentality is not a little thing as some would think, but a dry rot that destroys religion at its roots. One of the most common and dangerous forms of sentimentality that fastens itself on Christianity is what we shall call the evocative use of words.

SENTIMENTALITY

Words may be used either to designate an object or evoke a sentiment. In the evocative[1] use, when it becomes sentimental, the word is used for the sake of the sentiment attached to the word rather than for the purpose of designating some object. In ordinary language the word ought properly to fade out of sight as soon as it has served its proper function of referring to some object. This proper use of words we shall call the designative, to distinguish it from the evocative. In the designative usage all sentiment attached to the word is avoided as much as possible in order that any sentiment, if there be sentiment, shall attach itself to the object designated. There is, of course, an aesthetic use

1 This distinction between evocative and designative use of words is developed by Ogden and Richards in their *Meaning of Meaning*, which is an excellent analysis of language as a tool of thought.

of words in which the sound and subtle associations of the words are of more importance than
what it designates.   This is quite proper in poetry
and music.   But it becomes dishonest, and in matters of great practical importance, disastrous, to
pretend to use words for designating objects when
one is really using them for the sake of their sentimental associations.   For instance the word "the
Cross" is very commonly used by Christian people.
The critical question is: Does the glow of sentiment aroused by that word attach itself to some
specific object to which the word refers, or does it
attach itself to the word itself?   If the former, we
would ask just what is that specific object?   Is it
two pieces of timber attached to one another cross-
wise?   If so, then we claim that the word is used
evocatively, because two pieces of timber are not
proper objects of any such glow of sentiment or
high valuation, and this value would never attach
to pieces of timber were it not for irrelevant associations with "mother's knee" and a "picture on
the wall" and the days of childhood, and the
"little brown church in the vale" and the "prayer
at bedtime" and so on indefinitely and wearisomely.
We do not mean to imply that the word "the
Cross" has no specific object which it ought
properly to designate in Christian usage.   On the
contrary our claim is precisely that it does have
such an object and that therefore any one who uses
the word to designate two pieces of timber with
mawkish sentimentality is guilty of treachery to
the genuine and proper object of designation.

What has been said of "the Cross" would apply equally to a host of other words, such as "the Word," "the Blood," "the Spirit of Christ," "love," "brotherhood," etc., etc. These are dangerous words and should be rarely used if one reckons among the dangers of living the likelihood of becoming a traitor to the Christian faith through transferring loyalty from the proper objects of faith to the words that designate those objects.

Compare a group of engineers and a group of churchmen, respectively, each gathered together to discuss means of preventing, say, an impending war. All people are likely to fall into the sentimental use of words. But the engineers are much less likely to do so than churchmen, we believe. They are much more likely to use words to designate certain facts. To avoid ambiguities, and divest their language of those sentiments which so readily accrue to words in common usage, they often resort to technical terms, because technical terms are more precisely defined. Also they know the danger of using common words. They know that common words very readily gather irrelevant associations as a ship gathers barnacles. The word "Bolshevik" is an extreme example, few people using it to designate any specific object. But for engineers, speaking generally, it is the facts which engage their attention, rather than the words used to designate the facts. The churchmen, on the other hand, are prone to try to solve the problem by words that exhilarate more than illuminate. "When a public speaker has no clear view of the solution of his own problem, he always winds up by recommending the spirit of Christ. It never fails to bring salvos of applause. The people walk out with a rapture of exhilaration, believing they have actually got somebody out of trouble—the Armenians, or the citizens of Fiume, or some equally unhappy persons."[2]

Why is this treacherous weakness so deep and pervasive throughout prevalent Christianity? There are two reasons for it, we believe.

First, institutional Christianity has staked its success on the use of words. It has made talking its one preëminent agent of achievement. It measures its success by the number of people who will come out to hear a man talk. Now when talk is made to uphold such an enormous burden as this, it is bound to be twisted out of shape like steel girders before an avalanche. If crowds must be gathered and swayed by words, and if one man must do this twice a week or oftener, he is forced to make an evocative use of words. People will not sit and listen to words, and come in crowds to do so, unless, either the words designate certain facts of great interest, or else the words themselves thrill them, charm them, exhilarate them, bewitch them. But if it be the facts that interest them, they will not be content with listening to words about those facts. They will want to have more intimate converse with the facts. They will be content to listen to further words only occasionally, as these words may guide them into further and more inti-

2 Wiggam, A. E., *The New Decalogue of Science.*

mate experience of the facts. They will not be content merely to come regularly week by week to hear a man talk *about* these facts. It is not at all improbable that they might become so engaged with the facts as sometimes to be irregular in their attendance on the speaking of words about the facts. There should be some other standard for measuring the success of a church besides the number of regular attendants on the words.

A typical instance of this "Christian" faith in words rather than in God, came to the writer's attention some time ago. A speaker was addressing a group of students concerning the power which comes to the Christian by virtue of his religion and which must be the mark of a genuine Christian. He told of a young man who attained the highest honors in a great institution of learning. But when he came back to his home and attempted to speak in the cause of Christianity his speech stumbled; he could not hold the attention of his hearers. The moral drawn was that this young man had lost the power of the Christian faith. Evidently the power of the Christian faith was here identified with a glib tongue. It would seem that Christian power had nothing to do with the use of test tubes and statistical graphs; nothing to do with isolating the germ of influenza or discovering some such social device as the secret ballot by which more honest voting could be assured. Above all, it seemed to ignore that simple, unwordable devotion of friendship which is the most effective means of moulding human lives and is in no wise dependent upon glibness of speech.

A second reason why Christians are addicted to the evocative use of words is the obscurity of the objects with which Christian discourse is concerned. Scientific discourse refers to objects which can be clearly defined and very thoroughly investigated. Science is able to do this simply because it charts out a certain field of data which lie within the limits of its comprehension and investigation, and deliberately ignores everything which exceeds these narrow limits. It prunes down the objects to the limits of its own powers of definition and minute examination. But this Christianity cannot do. This religion cannot do by reason of the nature of religion. Religion must deal with the total concrete fact, even as love must. It must stretch the understanding to the dimensions of the Fact, not cut down the Fact to the dimensions of·the understanding. Hence words of religious discourse refer to objects not clearly definable. This opens the way for sentimentality in the use of words.

What we have just stated is an excuse for evocative use of words in Christian discourse, but it is not a justification. The obscurity of the object does not make inevitable the evocative use of words. On the contrary, words that refer to obscure objects can be just as designative as are scientific terms if the obscure objects are certainly known. Obscurity must not be confused with uncertainty. The objective, concrete existence of the religious object is certainly known to those acquainted with God, but he is obscurely known. When certainly known, the words used to refer to this concrete objective existence can be just as designative, if we are careful to make them so, as scientific words, although they cannot be as definitive of the object as are scientific terms in the exact sciences.

There is a third reason why Christian people become addicted to the sentimental use of words to the great harm of their religion. It is the requirements of religious ceremonial. The ceremony of public religious worship requires the evocative use of words. This is necessary if public worship is to be conducted properly and serve its rightful function of preparing the mind for the worship of God. Such being the case it is readily seen why those who have official charge of the conduct of such worship would become addicted to the evocative use of words in all matters that pertain to religion. Even when they are not conducting the ceremony of religious worship the habit persists. It is also natural that people who habitually participate in such worship should fall into the habit of using words in this way whenever they discuss matters in the name of religion.

The cure for this evil is not to abolish religious ceremonial. On the contrary, the ceremonies of public worship are highly valuable and must be developed far beyond their present efficiency; and such ceremonies must always use words evocatively. But the cure of the evil is carefully to guard against transferring the evocative use of words in ceremony where it is quite proper, into other departments and times and places where the evocative use is improper. The ceremonial and the expository departments of religious edification must be kept clearly distinct. Religious ceremony which does not designate facts, but only prepares the mind for worship by breaking down those practical and theoretical attitudes which render the mystic experience impossible, may properly use words evocatively, as lovers do. The ceremony must play upon the mind in such a way as to make it receptive to that unsifted mass of experience by which one perceives God. But when it comes to designating facts, as in expounding ideas about God, in presentation of practical problems or the like, words have a wholly different function. These two functions must be kept separate.

Let us see a little more clearly why ceremony must use words evocatively. Its function, as we have seen, is to prepare the mind to worship, to put the individual in that state of mind in which he is able to perceive God. That means that the jammed doors of the mind must be wedged open so that the fullness of experience may stream into the field of awareness and thus make God perceptible to the worshipper. There are two agencies which are best adapted to wedge open the doors of the mind. One is art and the other is love. Ceremony makes use of both of these.

It is very plain that religious ceremony is a work of art. There is music and rythmic intonation, there is harmony of color and motion. Even the

Puritan had his aesthetic demands for an austere and stately form of worship. But words cannot enter into a work of art unless they become evocative as well as designative. It is not their logical designation, but their sound and their associations, which make them acceptable to the aesthetic sense. Here, then, we have the first reason why evocative use of words in religious ceremony is inevitable. The aesthetic requirements of ceremony demand it.

The second agency which conducts the mind toward the mystic perception of God is love. Love, even more than art, enables the mind to enter into that fullness of sensory experience which is the nearest approach to worship. "He that loveth not his brother whom he hath seen, how can he love God whom he hath not seen." Among those who are gathered together for worship love must reign, and so it does, at least in theory. Ideally a group of worshippers is a beloved community. But love, rightly and of necessity, must use words evocatively. When a mother babbles to her child she plainly is not designating objects. She is simply evoking a sentiment in herself or in the child or in both. When two lovers are together and express themselves by such words as "darling" and "dearest" they are using words evocatively. Wherever words are used to express and evoke a sentiment, rather than to designate an object, they are used evocatively. In religious ceremony it is sought to express and evoke the sentiment of love. Hence the words used must be evocative.

The first requirement, then, if we are to keep

religious ceremony from contaminating the designative use of words in religious thought is frankly to admit the evocative nature of ceremonial words. This will keep us from the confusion which is so fatal.

A second requirement necessary to guard against sentimentality in language is to avoid using ceremonial words except when engaged in the ceremony itself. When, for instance, we are engaged in the discussion of religious matters in a theoretical or practical way, seeking to designate certain objects rather than cultivate the mystic state of worship, we should use other words, even when referring to the same objects as those with which the ceremonial words are associated. Also the words used to designate a certain object, especially those obscure objects which engage the attention of religion, should be constantly varied. If we constantly use the same word or phrase to refer to the same object the word will quickly gather to itself a sentiment of its own. Thus thought and sentiment will gradually become detached from the object and attached to the word. Thus we shall gradually betray the objects of religious veneration by detracting attention from them to the words used.

This use of words may block the way to God even for the isolated individual worshipper. When he repairs to worship he hits on a word or phrase or other symbol. What does it mean? Does it mean God? To mean is to refer to. But for Him these symbols do not refer directly to God but to some other words. What do these words mean?

Still other words. And so he goes on from words to still other words, not so much by logical definition as by idle association, until he hits on something which gives a rich glow of exhilaration because of its irrelevant associations. It may be a phrase taught him at his mother's knee or connected with something else of dear memory. It may be a bit of imagery of some sort. But in any case he stops with it and indulges in a gush of sentimentality. Then he comes away from his "worship" with a deep sense of sanctification, much soothed in spirit and very complacent. By persistency such a worshipper may so immerse himself in sentimentality as to become blind to all unpleasant facts about him, insensitive to every appeal, incapable and unwilling to lift a finger to help his fellow men, and able to gloat and fiddle while Rome burns.

Now such a "worshipper" has not found God at all. He has only found some symbol with rich associations attached to it. Such a symbol may be of great service in preparing the mind for worship. All worshippers, perhaps, make use of such symbols. But for the true worshipper they are means, not ends. They are not the objects referred to, but the instruments of reference which refer the worshipper on to the divine Object. To the true worshipper such a symbol is the medium through which he discerns God. But to the sentimentalists, it is an obstacle which blocks the way to God. He looks at it, rather than through it. He gloats upon it, rather than uses it.

This error of the sentimental worshipper is the evil of idle dreaming. It is an evil into which worshippers are often likely to fall. It is the idle following of fancy rather than the strenuous discernment of objective fact. It has often been confused with mysticism because mystics or pseudo-mystics have so often fallen into it. It has put mysticism into bad repute among many. Because of this idle dreaming the logical thinker and practical worker have scorned mysticism. But they have confused true mystic discernment with idle dreaming. They are right in denouncing what they denounce; but they fail to see that true mysticism does not properly come under their denunciation.

TRADITION

As scientific method serves to save Christianity from sentimentality, so also it may deliver religion from the bondage of tradition. Tradition has its rightful place in any religion, as we shall take pains to show, but religion is peculiarly liable to suffocation from an overgrowth of tradition uncritically accepted. Tradition perverts religion unless the critical mind is turned upon it; and the critical mind is science. As Francis Bacon so clearly announced, the first task of science is to turn upon tradition with criticism. The great fight of scientific method has always been against the inertia of tradition.

Mysticism, without the support of science, cannot deliver religion from the dead hand of tradition. On this matter no one has better spoken than

Dean Inge, himself a great mystic and expounder of mysticism.

> The mystic who refuses to analyze or criticise his intuitions is often baffled by the emptiness or formlessness of his religious conceptions, and so tends to fall back upon the clearly defined images or symbols which his church provides. He accepts these on authority, since he is not interested in the proofs of them, and would even value them less if they were based on ordinary evidence. Whether consciously or not, he only needs them as helps to his imagination. But they may easily become so indispensable to him that he will be as stiff a dogmatist as if his Faith really rested on external authority; and he will often protest vehemently that external authority, in the form of supernaturally revealed doctrines, is in truth the basis of his Faith, which would fall in ruins if this support were withdrawn. Just because the dogmas of his church are accepted uncritically, as outside discussion, they are capable of being used as external suports of a Faith which in reality sprang up independently of them, and only requires them to give form and colour to its vague intuitions. The typical dogmatist is a confused half-mystic, whose intuitive Faith is neither strong enough nor clear enough to bring him strength or comfort. He accordingly fortifies himself by calling in the help of an external authority, whose credentials he would think it impious to investigate, and willingly accepts its guidance whenever the inner light burns dim.[3]

---

3 *Faith and Its Psychology*, p. 82.

This combination of mystic and uncritical traditionalist, makes the worst kind of bigot. He is the fiery defender of the faith who cannot rest nor let any others rest and who will sacrifice his own happiness and the happiness of those that are dear to him. He will not compromise nor yield on any point. He is a fanatic. He is genuinely sincere. We cannot help but respect him if we know him well. He is a victim of that tradition which gives "form and colour to his vague intuitions." He is a victim quite as much as those whom he persecutes. It is the business of science, when properly adjusted to religion, to save such victims and to deliver religious faith from such perversions. There has been far, far too much of it in the history of all religions.

But religion cannot dispense with tradition. Neither can science, for that matter. It is not a matter of discarding tradition; it is a matter of interpreting and criticizing tradition. Christianity is peculiarly dependent upon tradition.

Christianity in a very broad sense consists of teachings and sentiments which have been developed throughout many centuries by innumerable men, and transmitted to the individual by tradition, so that he comes to his immediate experience of God equipped with these ideas and sentiments for interpreting his experience and discerning its significance. Thus only can he have Christian acquaintance with God.

Christianity in a somewhat narrower sense

would be those teachings and sentiments which would be developed if we could limit the source from which they were derived solely to the experience and utterances recorded in the Old and New Testament. But as a fact this is impossible, for the teachings of the Bible come down to us through the transmitting medium of the thought and life of many generations of men, and especially through the medium of the thought of our own age. This medium gives color and form to whatever meaning we derive from the Bible, and any one who thinks he can escape the influence of this medium through which he reads the Bible, simply puts himself in the narrow bondage of some special tradition which probably distorts the original meaning of the book more than most mediums.

Christianity, in still a narrower sense, would be those sentiments and teachings about God which would arise if we could limit their source solely to the life of Jesus Christ. Needless to say this is even farther removed from our actual Christianity than that described in the preceding paragraph.

Christian worship, then, as found in the world today, consists of two parts: The immediate experience of God through mysticism; and the significance of this experience as revealed by the teachings and sentiments handed down to the individual by way of Christian tradition, and preëminently through the Christian church. These traditions are constantly subject to correction by study of the Bible and especially study of the life and teachings of Jesus. But these teachings and sentiments do not of themselves yield acquaintance with God. They yield only knowledge about God. They can enter into one's acquaintance with God only when they are used to reveal the significance of that immediate experience which one attains in mystic worship. The God-acquainted Christian must, then, have two things: (1) the immediate experience; and (2) the teachings which interpret that experience. And no man is truly a Christian unless he has both. Only then does he belong to the God-acquainted.

Let us quote again from Dean Inge, this time concerning the rightful place of tradition or, as he prefers to call it, authority, in religion. "The office of authority in religion is essentially educational. Like every good teacher, it should labour to make itself superfluous. The instructor should not rest content till his pupil says, 'Now I believe not on thy saying but because I see and know for myself.'" We do not believe that authority can ever become superfluous if it is identified with tradition. It can shape and guide the original experience of the worshipper, and the worshipper can shape and reinterpret it, so that the two fuse into one thought by which God is known. But with the wealth of tradition at our disposal and with any growing wealth of original experience, the time is never reached when some further readjustment is not required between authority and immediate experience of God.

And it is in this work of readjustment that scientific method has its part to play. Tradition can be analyzed, reinterpreted, and properly applied to the interpretation of original experience only by those methods of accurate thinking which go by the name of science.

CHAPTER III

THOUGHT AND WORSHIP

In all that we have said of religion's needs of science we do not want to give the impression that we think religion can ever become a science. It can never become a science any more than love or aesthetic appreciation can. A correct understanding of the object is only one ingredient that enters into worship, love and aesthetic appreciation. In fact these three can develop to a high degree with very little correct knowledge of the object. And yet these three almost inevitably take on pathological forms if illusions continue to develop without correction. Cognition of the object is one ingredient, even though it be but one. Science has its contribution to make to the welfare of this side of life, although it can never be substituted for this side. Religion, because of its essential nature, is eternally to be distinguished from scientific knowledge pure and simple. Worship must absorb science, but it must go on beyond science. It must not contravene the verified conclusions of scientific method, but it must not be narrowed down to scientific knowledge pure and simple. For it is more than knowledge. It is love.

We must try to show how scientific thought can be taken up into worship, love and aesthetic appreciation; how these latter can transform scientific knowledge without contravening it. Science is cognition in its more rigorous form; and we want this most rigorous form of cognition in our apprehension of our beloved and of our God. But such scientific cognition by itself alone can never satisfy us in dealing with that which we love and worship. How can we be true to science and yet exceed science? That is our problem.

To indicate the way in which love and worship may incorporate and yet exceed the conclusions of scientific method we shall adopt the word contemplation. We shall use it in a slightly different sense than its ordinary usage allows, and yet we believe we shall not depart altogether from its common meaning. We believe it is sometimes used in our sense.

The chief objection to the term contemplation is that it seems to connote passivity. So it does ofttimes, but not necessarily. It is possible for most strenuous endeavor to be contemplative. An infant is contemplating an object when he is squeezing it, rubbing, licking, sucking, pressing, poking, prying, dropping, picking up again, etc. He is endeavoring to experience it completely, to take it all in, to draw from it all the qualities it is capable of yielding, and to re-experience those qualities until they become familiar. He is striving to cognize the total, concrete object. The enthusiastic mechanician who has a passion for machinery is contemplating when he works upon it, however practical his achievement. He delights in repairing, constructing, and reconstructing machinery because in that way he is able to react to it in its totality. The concrete totality of the mechanism, all its parts and their interrelations, can be most completely grasped by him and become, as a concrete whole, the object of his response, only as he thus works with it. Hence his work with machinery is joyous work. The same is true of a mother caring for her infant. She may make use of the child, but the child is more than a utility. She may theorize about the child, but the child is more to her than merely an object on which to experiment to the end of testing and developing her concepts. It is a concrete object which she endeavors to comprehend in its totality. She may be most strenuously active in her care of the child, and this activity may yield the satisfaction of any primitive instinct. But it differs from the biological instincts as the play of the child differs from the play of lower animals. In both cases the humans, both the mother in care of the infant and the child at play, seem to be actuated by the endeavor to comprehend the concrete totality of the object, while the lower animals only go through a more or less fixed repertoire of actions. The puppy never plays he is a little man, but the child does play he is a little dog and much else besides, for in so doing he amplifies his experience and comprehends more of that with which he deals. So the mother in her care of the infant labors to "take it all in" and finds delight in all her work with it because she so fulfills this desire. So likewise we might make a study of all forms of joyous work. Always we find that the contemplative element enters in.

There is, however, justification for the prevalent view that contemplation is a passive state, for it often takes that form. But the passivity is only due to our limited capacity to appreciate any more than what has already been experienced. Hence we cease to extend the range of our experience by further activity and give ourselves over to reviewing and appreciating that which is already accessible to our minds from past experience and immediate sensations. Additional experience would only disturb and hinder our comprehension of the total object. Our capacity for contemplation has certain limits and when we try to go beyond these, contemplation breaks down and our mental attitude reverts to the narrow bounds of the practical and theoretical. To avoid this, contemplation guards itself by assuming the passive state. But even when thus passive so far as overt behavior is concerned, we are perhaps responsive to many more stimuli when contemplating than when engaged in a purely practical theoretical undertaking.

The chief significance of contemplation for us just now is that it represents that kind of cognition which may be highly scientific and yet is more than science. It is the form in which thought and intellectual rigor can enter into religion. In contemplation worship can absorb all that science has to offer and yet exceed science. To make this plain let us contrast the three mental attitudes we have already referred to several times, the practical, the theoretical, and the contemplative.

There are some objects which we cognize only to that degree and in that way which will enable us to use them for the satisfying of some urgent need. This is the practical attitude.   Other objects we note only with respect to those features which are relevant to some theory, to the end of testing and reshaping the theory or classifying the object. This is the theoretical type.   In both practical and theoretical attitudes we ignore most of the experience which pertains to the object.   In the contemplative attitude, however, we are receptive to all the experience which seems to pertain to the object. We may cognize a child in the practical attitude. We study how to deal with him so as to get him to do our will with the least sound and fury.   Or, being a student of genetic psychology, we consider him with a view to testing some theory.   But if we love the child we may sometimes contemplate him.   In so doing we may keep true to all that practice and theory has revealed about him; but our attitude toward him is very different and what we discern in him far exceeds what enters our minds as we theorize or practically deal with him. Far more of the concrete total individuality of the child comes to our knowledge when we contemplate him.

Contemplation is scientific but much more than science, because it involves what approximates the response of the total personality to the concrete object, while in the practical and theoretical attitudes only fractions of the personality respond to certain selected features of the object.

Thinking dominated by the practical attitude is most common. Men ordinarily do not like to think, but do so only when in trouble and in order to get out. They think to rid themselves of annoyance, protect themselves from danger, reach a place of security, or free themselves from the need of further thinking. Such thinking is a striving to escape from the difficult situation in which thinking is required into that where automatic habits are sufficient to meet all demands. It is an attempt to adjust habits in such a way that they can operate without the need of further thought. We do not think as long as the street car runs according to schedule and we can get downtown merely by following our automatic habits of walking, riding, etc. But when the street car ceases to run, we begin to think how we can get downtown. As long as our feet carry us automatically, whither we will, we do not think about our walking. But when something obstructs our path, when our feet stick or begin to sink, our walking comes under the direction of thinking. Instrumental thinking occurs whenever our established habits are disturbed and our sole motive is to regain the mindless automatism of the habitual routine. In instrumental thinking we never rise out of the old ruts except when forced out, and then our thinking is solely for the purpose of getting into the old rut, or else developing a new rut as narrow and confining as the old one. Instrumental thinking strives to get into a situation where thinking will no longer be necessary.

But under the dominance of the theoretical attitude thinking may change its complexion quite completely and become a form of play. Because of the pleasure which some people find in mental gymnastics, we might call this hedonistic thinking. It may take the form of argumentation upon almost any subject whatsoever. Interest in Chinese puzzles, chess, riddles, invention of queer devices, and the satisfaction of idle curiosity, in almost any field, all display it. William James describes this motive to thought by the term "The Sentiment of Rationality." The delight of hedonistic thinking is the pleasure of finding some abstract order running through the countless details of the concrete world. The Chinese puzzle is a wild confusion until one "gets the hang" of it, and then it resolves itself into a beautiful order. The same is true of the mathematical problem, of the chess game, or the invention of a device out of a heterogeneous aggregation of things. For we must understand that mechanical invention by no means always has the practical motive behind it. For the scientist, engaged in what is called pure science as distinguished from applied, the universe, or rather some selected portion of it, is a Chinese puzzle in which he seeks to discover some order. According to his starting point he may discover many different orders for his delight. Philosophers share this motive to no less degree. Josiah Royce and Bertrand Russell have both confessed to this hedonism of thought, finding in it a pleasure without regard to any use it might have. The interminable maze of intellectual operations we sometimes find in philosophic, mathematical and scientific endeavors plainly could never have been developed had there not been a strong hedonistic tone in the mental operations themselves.

We wish to say nothing in disparagement of such a motive to thought, nor of such thinking. Some of the greatest services to mankind have come out of the results of such thinking. The very usefulness of science consists to a large degree in freeing the scientist from the constraint of immediate, practical needs. He must wander far and wide into unknown territory, not knowing what he shall find. Others besides Saul have found a kingdom by accident. But it is plain that such thinking is not religious. It is not of necessity anti-religious. Furthermore, the products of such thinking can be taken up into religion just as they may be taken up into practical life. But one cannot investigate God or the religious experience in this spirit and at the same time be in a worshipful attitude. Indeed if this were the only way one cognized any object of experience he would never know God as the religious person knows Him. The genetic psychologist can never know the child as the parent knows him, unless the psychologist can also love and contemplate. The geneteic psychologist may assist the parent to know his own child, but it is the parent and not the psychologist who will know the child most profoundly, providing the parent is willing to learn from the psychologist. And this will not be due to the fact that the parent has more constant association with the child and so will learn much from trial and error. We refer to something very different. We refer to that receptivity to the concrete experience of child which goes only with the contemplative attitude. We refer, furthermore, to that response of innumerable impulses in the parent to the child, which are not awakened in the psychologist. All this wealth of experience and response in the parent will not necessarily yield correct knowledge. The parent may be subject to most grotesque illusions concerning the child. But if the parent is willing to learn from the psychologist, to have his errors corrected and his ideas clarified, then he can know the child as the psychologist never can.

Now all this applies to science and religion. The theoretical attitude is not the worshipful attitude and never can be. We can never cognize God by theoretical thinking alone, and have any religion. But if science and religion were rightly adjusted to one another, the worshipper of God could profit as much by science as the loving parent can profit by the genetic psychologist.

In contemplation the skeleton of scientific theory is clothed with a mass of concrete experience which makes the object very much more than what it is described to be by the scientific theory. The scientific theory may be correct, but it falls far short of the total object as cognized in contemplation.

The glowing eye, the yellow curls, the ruddy cheek, but more important still, all that which we have experienced in the child in the past which makes up our "apperceptive mass" as we contemplate him —the deeds and words, and triumphs and sorrows, all that has entered into his life in so far as it is preserved in our minds as we look upon him—all this we behold in the child.   And all of this, or at any rate much of it, is fact.   It is that which we know.

In the same way also we may know God.   When the total Object of all our experience becomes for us an object of contemplation, we discern that which stirs us deeply, that which awakens in us the religious attitude.   There flows in upon us masses of experience, somewhat after the fashion that experience rolls in upon us in contemplating the child; but in the case of cognizing God the experiences do not center about one single human organism, but may be inclusive of almost anything and everything we have ever experienced.   When we thus contemplate the total Object of experience, we have the religious experience.

But no matter which motive may dominate our cognition even though it be the contemplative, it is plain that there is a great waste of experience. For our power of taking in experience, in the sense of finding any interest or significance in it, is very limited.   There is a vast dump heap of experience, mounting to infinity, which we completely ignore. The trivial, irrelevant, nonsensical, useless, accidental, and meaningless elements of experience make a huge rubbish heap which constitutes the greater part of this universe, if we may call the totality of experience a universe. The significant, useful, and interesting bits which we select constitute only the tiniest structure in comparison.

But are these trivial, uninteresting, insignificant, and useless elements of experience which throng in upon us all the time absolutely so? Of course not. They are trivial and uninteresting only because we have not discovered those relations in which they are important and interesting. We have not found that Object of contemplation in which they can be integrated. We have not gotten that slant or angle of approach from which their light becomes apparent. They are irrelevant only because our purpose is so narrow as to render them irrelevant to it. They are useless only because our enterprise is not big enough to make use of them. Human life at its best might be described as the search for that total Object or bigger Fact in which all experience might be integrated and made to yield up its maximum significance. This search finds its culmination in religion. For this reason, also, it is plain that God is necessarily that Object, however unknown, which must bring human life to maximum abundance when man makes proper adjustment to Him.

Now contemplation is precisely the thinking which is dominated by the motive to integrate experience more widely and completely. Of course the actual amount of experience which humans have thus far been able to integrate and interpret as compared to the total mass of experience, is very tiny indeed. But contemplation is the thinking that drives in that direction. It seeks to redeem the waste places. It is the endeavor to discover some system of integration which will draw into the object of concern more of that which otherwise is irrelevant and insignificant. It endeavors to find a method of fitting bits of experience together so that more of the offscourings can be turned to use. The great religious quest is for that total Object which would render all experience significant and so bring human life to its highest pitch of enrichment and interest.

CONTEMPLATION AND MYSTICISM

Mysticism in its more extreme forms is not a knowing state at all. It is a form of immediate experience, while contemplation is a form of cognition. Contemplation is midway between mysticism—where wealth of experience excludes cognition—and practice-theory, where cognition excludes wealth of experience. In contemplation we have some well-defined beliefs concerning the object of experience. In mysticism beliefs and all forms of cognition fade away to the minimum and the mind resolves into a state approximating simple awareness.

There are certain legitimate reasons why exponents of religion have so frequently turned away from mysticism as the source and means to religious certainty. The chief of these has been failure to make the very distinction we have just noted. The mystic experience has been identified with the truth; but the truth can only be the correct meaning of that experience and not the bare uninterpreted experience itself.

There has been another difficulty. It has been confusion between the datum of the mystic experience and the individual mystic's personal reactions to that datum. In other words, we have failed to distinguish between the raw, uninterpreted datum of experience given in the mystic state and the visions, photisms, convulsions, involuntary vocal utterances, descriptive symbols, philosophical deductions, theological ejaculations, and other personal idiosyncrasies which have characterized the individual undergoing that experience. All this confused mass of personal reaction and idiosyncrasy which has overgrown, and sometimes almost concealed, the original datum of experience, must be cleared away in order to give us access to the immediate experience itself. This overgrowth reveals the psychic constitution of the individual mystic rather than the object or meaning of his experience.

Let us turn again to William James for descriptions of mystic experience as sources of insight. In his essay "On a Certain Blindness in Human Beings," taken from *Talks to Teachers,* we find record of these cases.

> Richard Jeffries has written a remarkable autobiographical document entitled *The Story of My Heart.* It tells in many pages of the rapture with which in youth the sense of the life of nature filled him. "On a certain hilltop," he says, "I was utterly alone with the sun and the earth.

Lying down on the grass, I spoke in my soul to the earth, the sun, the air, and the distant sea, far beyond sight. . . . With all the intensity of feeling which exalted me, all the intense communion I held with the earth, the sun and sky, the stars hidden by the light, with the ocean—in no manner can the thrilling depth of these feelings be written—with these I played as if they were the keys of an instrumnt. . . . The great sun, burning with light, the strong earth—dear earth—the warm sky, the pure air, the thought of ocean, the inexpressible beauty of all filled me with a rapture, an ecstasy, an inflatus. With this inflatus, too, I prayed. . . . The prayer, this soul-emotion, was in itself not for an object; it was a passion. I hid my face in the grass. I was wholly prostrated, I lost myself in the wrestle, I was rapt and carried away. . . . Had any shepherd accidentally seen me lying on the turf he would only have thought I was resting a few minutes. I made no outward show. Who could have imagined the whirlwind of passion that was going on in me as I reclined there!"

Surely, a worthless hour of life when measured by the usual standards of commercial value. Yet in what other kind of value can the preciousness of any hour, made precious by any standard consist, if it consist not in feelings of excited significance like these, engendered in some one, by what the hour contains?

Yet so blind and dead does the clamour of our own practical interests make us to all other things that it seems almost as if it were necessary to become worthless as a practical being, if one is to hope to attain to any breadth of insight into the impersonal world of worths as such, to have any perception of life's meaning on a large objective scale.

James quotes from Walt Whitman. Those of us who know him need only the mention of his name to recall how marvelous was the scope and vividness of his response to all the sights and sounds of the world. The gates of his sensitivity were as wide open as ever were any man's, it would seem, to all the concrete world. Typical is a letter by Whitman, which James quotes.

Dear Pete:—. . . . You know it is a never-ending amusement and study and recreation for me to ride a couple of hours on a pleasant afternoon on a Broadway stage in this way. You see everything as you pass, a sort of living endless panorama—shops and splendid buildings and great windows; on the broad sidewalks crowds of women richly dressed, continually passing, altogether different, superior in style and looks from any to be seen anywhere else—in fact, a perfect stream of people—men, too, dressed in high style and plenty of foreigners—and then in the streets the thick crowd of carriages, stages, carts, hotel and private coaches, and in fact all sorts of vehicles and many first-class teams, mile after mile, and the splendour of such a great street and so many tall, ornamental, noble buildings, many of them of white marble, and the gaiety and motion on every side; you will not wonder how much attraction all this is on a fine day to a great loafer like me, who enjoys so much seeing the busy world move by him and exhibiting itself for his amusement while he takes it easy and just looks on and observes.

Further on James adds:

When your ordinary Brooklynite or New Yorker, leading a life replete with too much luxury, or tired and careworn about his personal affairs, crosses the ferry or goes up Broadway, his fancy does not thus "soar away into the colors of the sunset" as did Whitman's, nor does he inwardly realize at all the indisputable fact that this world never did anywhere or at any time contain more of essential divinity, or of eternal meaning, than is embodied in the fields of vision over which his eyes so carelessly pass. There is life; there, a step away, is death. There is the only kind of beauty there ever was. . . . But to the jaded and unquickened eye it is all dead and common, pure vulgarism flatness and disgust.

Life is always worth living, if one have such responsive sensibilities. But we of the highly educated classes (so-called) have most of us got far, far away from Nature. We are trained to seek the choice, the rare, the exquisite, exclusively, and to overlook the common. We are stuffed with abstract conceptions and glib with verbalities and verbosities; and in the culture of these higher functions the peculiar sources of joy connected with our simpler functions often dry up, and we grow stone-blind and insensible to life's more elementary and general goods and joys. . . .

The savages and children of nature, to whom we deem ourselves so much superior, certainly are alive where we are often dead, along these lines, and, could they write as glibly as we do, they

would read us impressive lectures on our impa-
tience for improvement and on our blindness to
the fundamental static goods of life.   "Ah! my
brother," said a chieftain to his white guest,
"thou wilt never know the happiness of both
thinking of nothing and doing nothing.   This,
next to sleep, is the most enchanting of all things.
Thus we were before our birth, and thus we shall
be after death.   Thy people . . . . when they
have finished reaping one field, they begin to
plough another; and, if the day were not enough,
I have seen them plough by moonlight.   What
is their life to ours—the life that is as naught
to them?   Blind that they are, they lose it all!
But we live in the present."

The intense interest that life can assume when
brought down to the non-thinking level, the level
of pure sensorial perception has been beautifully
described by a man who can write, Mr. W. H.
Hudson, in his volume, *Idle Days in Patagonia.*

Not once nor twice nor thrice, but day after
day I returned to this solitude, going to it in the
morning as if to attend a festival, and leaving it
only when hunger and thirst and the westering
sun compelled me.   And yet I had no object in
going, no motive which could be put into words;
for, although I carried a gun, there was nothing
to shoot—the shooting was all left behind in the
valley. . . . Sometimes I would pass a whole
day without seeing one mammal, and perhaps not
more than a dozen birds of any size.   The
weather at that time was cheerless, generally with
a gray film of cloud spread over the sky, and a

bleak wind, often cold enough to make my bridle-hand numb. . . . At a slow pace, which would have seemed intolerable under other circumstances, I would ride about for hours together at a stretch. On arriving at a hill, I would slowly ride to its summit, and stand there to survey the prospect. On every side it stretched away in great undulations, wild and irregular. How gray it all was! Hardly less so near at hand than on the haze-wrapped horizon where the hills were dim and the outline obscured by distance. Descending from my outlook, I would take up my aimless wanderings again and visit other elevations to gaze on the same landscape from another point; and so on for hours. . . .

In the state of mind I was in, thought had become impossible. My state was one of suspense and watchfulness; yet I had no expectation of meeting an adventure, and felt as free from apprehension as I feel now while sitting in a room in London. The state seemed familiar rather than strange, and accompanied by a strong feeling of elation; and I did not know that something had come between me and my intellect until I returned to my former self—to thinking and the old insipid existence (again).

I had undoubtedly gone back; and that state of intense watchfulness or alertness, rather, with suspension of the higher intellectual faculties, represented the mental state of the pure savage. He thinks little, reasons little, having a surer guide in his (mere sensory perceptions). He is in perfect harmony with nature, and is nearly on a level, mentally, with the wild animals he preys on, and which in their turn sometimes prey on him.

We believe these quotations set forth a very important truth. But we believe there is also implied a grave error in the last quotation, and perhaps running through the others. We do not believe it true that the savage and the lower animals are responsive to a greater number of stimuli than the civilized man. It is not true that they are more alert to the fullness of sense data, necessarily, because of a "suspension of the higher faculties." As we have already said, when the higher intellectual faculties are suspended, we have something else in operation which may reduce the range of susceptibility even more narrowly, and that is routine habit. The savage and the lower animals are dominated by these habits (or instincts). They may seem to respond to a wider range of stimuli, because they react to different stimuli, than we. That which we ignore, they receive; but also much that we react to, they ignore. Hence the range is not necessarily wider.

But apart from these possible faulty interpretations and applications, the facts recounted in the quotations still stand. There are times when men, with a partial suspension of the thought processes, become blissfully athrill with the vast fullness of sensuous experience that rains down upon them. As soon as they begin to "think" the blinds are drawn down, the walls rise around them, and they cease to be alive with respect to all this fullness of the immediate world that encompasses them. All the world around them becomes dark and dead, as though it were not, save only that little streak across the surface of things, which is illumined by thought, because it happens to be that which you "are thinking about." All the rest is swallowed up in darkness just as the waters at night over which your ship glides are illumined only in those streaks which waver across from the scattered lights upon the shore. So it is that thought closes the mind and draws narrow limits about that small portion of the world with which you converse.

That kind of thinking which is most receptive to what mysticism has to offer is the contemplative. If mysticism ends with itself it amounts to little. Its value is that it opens up new undefined reaches of experienced reality. But if these new regions are not entered and possessed by thought, human life is not greatly enriched. Contemplation is best fitted to enter in and possess the land. The swing of the pendulum of interest from mysticism back to scientific method and from scientific method to mysticism, is of value only as it serves in each swing to build up a little more the breadth and fullness of that which we contemplate. And contemplation culminates in the discernment of God. Worship at its best is that contemplation which is finely balanced between thinking and mysticism, and fulfills itself in action.

Only by developing a scientific technique which is fit and able to interpret correctly the significance of that which is given in immediate experience, when immediate experience is at that floodtide called mysticism, can God be known. It is probable He can never be completely known; but we can increase our knowledge of Him by contemplation, which draws on mysticism from the one side and scientific method from the other.

CHAPTER IV

CHRISTIANITY AND LOVE

The contemplative way of life which we have been trying to describe includes, on the one hand, maximum appreciation and awareness of sensuous experience, and, on the other, the largest practical achievement and intellectual rigor. These two sides of life, the appreciative and the efficient, it unifies in such a way that each promotes the other. It is a dynamic, creative way of living. We call it contemplative because we have no other term to designate it. We would prefer another term if we had one, because contemplation is so commonly associated with passivity.

Now this way of life, which is both active and appreciative, intellectually accurate but at the same time receptive to the concrete fullness of sense, this way of life appears most completely in love. Indeed there is no other form known to man in which this contemplative life can be developed to such a high degree. Love yields the most full-orbed life precisely because it does bring about this unification of opposites and provides that delicate and difficult balance which we have been tracing throughout the foregoing chapters; in love at its best we find on the one hand that striving for knowledge freed of illusion which culminates in scientific method, and on the other hand that striving for fullest immediate awareness of fact which culminates in mysticism. In love, then, we find reconciliation and unification of that which distinguishes science on the one hand and religion on the other.

It may be scarcely correct to say that in love we find science, for science is a term used to refer to a mental attitude and method which must, when in actual operation, ignore that wide open awareness of experience which characterizes love, religion and aesthetic appreciation. But we can say that love unifies science and religion, the efficient and the appreciative, the abstract part and the concrete whole theory and belief, the verified proposition and sensuous awareness, because in love the truths discovered by scientific method can be merged and bodied out by the rich awareness of immediate experience. And there is no other way of life in which this can be so completely and satisfactorily done as in the life of love. To make this clear we must clarify the concept of love.

As an illustration of love at its best let us take the statement—that has fallen into our hands—of one who is surely a saint. We have authoritative information from those who know him well that what he here portrays is not a theory merely, but his own actual, constant way of life. It is his confession. He describes this way of love and religion in the third person, but he is speaking out of his own experience. Here we believe is to be found the scientific method merged with mysticism and the two interacting creatively. Note particularly his statement that one must "lay himself open before the richest heritage of the race, and counsel and criticism he will seek and weigh carefully. But at last it is to his own inner light that he must be true."

When it comes to guidance for the actual steps he must take each day, he will turn to the God within himself—that Love in every human heart which reaches out in a supreme concern for the good of all men alike and which will never fail to make itself known to any man whomsoever who will do two things: (1) go apart into a quiet place by himself and there relax all sense of strain and hurry, with the single purpose to listen for It; and (2) lay aside every desire or anxiety for himself, waiting (if necessary for months or years) until he be entirely clear that the Voice or Intuition within him is no other than that Universal Love which ever waits each man's surrender to It. In this experience he will find his anchor-hold on God. For him, thenceforward, there can be no unquestioning obedience to another's commands nor submissive conformity to orthodoxy or convention. He will lay himself open before the richest spiritual heritage of the race, and counsel and criticism he will seek and weigh carefully. But at last it is to his own Inner Light that he must be true. Nowhere in the world is there a code of morals, or universal law of right and wrong, that can be applied to all alike. Every man's law is within himself, waiting for his discovery of it. For each man it is different and yet for all alike it has the same meaning. It is that of growth toward the perfect goodness of God. . . .

His, then, are the freedom, the joy, the life, the wealth of the universe. Heaven is wherever God is. God is wherever love is. And if love wholly possesses a man now, for him Heaven is here— the "impossible" utopia of the Community of Love has come. And nothing can take it away from him—neither poverty, nor prison, nor "failure," nor death. For he knows that there is no poverty for him who is rich within, nor prison for him who is free within, nor failure for him who will not bend the knee to wrong, nor death for him who has found the real Life.

Here is Christianity in terms of love, and we know of no better terms in which to express it. But before we can adopt such terms we must distinguish clearly between genuine love and sentimental love. Sentimental love is blind. It feeds on illusion. It is a way of dreaming; it is not a way of cognizing. But genuine love is a manner of cognition. It is that type of cognition which in the previous chapter we described as the contemplative.

Sentimental love is so common, not merely between the sexes but in all human relations, that its blindness and illusions have been thought by many to be a characteristic of all love. Sentimental blindness is not an excellence but a fault, and while perhaps always present in human love to some degree, becomes less and less as human love becomes more excellent. Love is sentimental, blind and false, just in so far as any characteristic of the beloved, whether a virtue or a fault, is ignored.

Being blind to all faults in the loved one, sentimental love cannot do any thing to correct faults. Neither can it do anything to strengthen the virtues of the beloved because it is blind to the facts and cannot discern the internal obstructions with which the beloved must struggle. Sentimental love merely luxuriates in its own dream of an abstract, perfect object which does not exist anywhere in reality and probably would not be lovable if it did exist. Sentimental love in this way is very selfish. It would be selfish even though the sentimental lover gave up his life for his dream.

Sentimental love is so widespread and so commonly glorified as the noblest kind of love that it is important to point out some of its greatest evils and negative qualities.

We have said that it is unfitted by its nature to help the beloved one attain his own largest happiness and fullest self-development. It is a millstone about his neck because it tends to fill him with an unjustified complacency and to blind him to the facts about himself and the world around him. To be blind in a rushing, whirling world of hard realities is exceedingly dangerous and in the end is bound to bring suffering. Sentimental love transmits its blindness to the one who is loved. It prevents him from struggling with those real difficulties by which alone he can attain an abundant life.

It is a form of unconscious hypocrisy on the part of the lover. He is deceiving himself and loving his own self-deceit. This self-deceit tends to spread out into other branches of his life. All reactions to the true individuality of the beloved are suppressed. The lover uses his beloved very much as a little girl uses her old rag doll, as a bit of matter about which to build her fancies. But there is no evil in the little girl's play because she has no established illusions concerning the reality of the rag doll. Or, if she does suffer some illusion, it does not tyrranize over her. She can easily escape it. And in any case the illusions of childhood are harmless or excusable as a process of growth. They are sloughed off with maturity. But in maturity one must put away childish things. In sentimental love one is not playing with a rag doll, which is inert and without feeling, but with a live, human person. And this make-believe is made to function in the grim process of actual living. On it is made to depend, unlike the play with rag dolls, the fortunes of self, the beloved and society. When one begins to construct a skyscraper on foundations of make-believe, danger is ahead. Sentimental love is not merely illusion, but illusion which is made to serve as foundation for real life-building.

In sentimental love there is not community of mind. There is no mutual understanding. There cannot be when one knows not a real person, but only a make-believe person with a corporeal object to serve as a rag doll symbol. Yet the glory and blessing of love is very largely just this blending and interaction of visions by which arises a larger, fuller discernment of fact shared by both.

Sentimental love is competitive, which means that the more we love one, the less we can love others. This follows naturally from its exclusive nature. The sentimental affection we give to one is subtracted from the world at large. The energy and devotion we give to an unreal dream always diverts our energy and devotion from facts. Modern psychology has made us well acquainted with these dream substitutes for actual living. Sentimental love is an example of such a dream substitute and it has the same effect of diverting the energies from the real world and so from the real people that live there. We find mothers who love their children in this exclusive, competitive fashion. Other parents, on the other hand, find that affection for their own children quickens affection for all children because, when it is a real child that is loved and not merely a dream child, they find that the loved child is bound up with all other children. Any part of the actual world is bound up with other parts, hence genuine love for any part leads on to affection for other parts and the whole world, generally speaking, becomes dearer through genuine love. In sentimental love the opposite of this is true.

For the same reason sentimental love is non-progressive. The dream does not lead on to more and more continuously, because it is not an integral part of the inexhaustible and organic totality of the world. The dream soon reaches the limits of its elaboration and turns back upon itself to revolve round and round in a narrow circle of fancy. In the course of time a dream wears itself threadbare.

Sentimental love is rarely permanent because human nature is prone to rebel in the course of time against continuous self-deceit and suppression of all reactions to fact. In association with the beloved his true qualities must inevitably be apprehended in part, howbeit subconsciously. Reactions to the real individual will be aroused but not admitted to consciousness. In the course of time these reactions may accumulate and form a system of response, which might be called a complex, which runs counter to the response which has the dream as its object. Thus arises an inner conflict in the lover between the suppressed complex of criticism or even repulsion, and the response of sentimental love. This conflict may manifest itself in many different ways. It may show itself in nervousness and depression which have no manifest causes. In extreme cases it may lead to mental aberration. These conflicts due to sentimental love are one of the common sources of mental disorders. Or again, the conflict may show itself in alternations between honeyed endearments and violent quarrels. The suppressed criticisms break forth from time to time in bitter invective, then are suppressed and sentimental love again flows placidly on. Still again, the conflict shows itself in an undefined discontent. The lover cannot understand why he is not completely happy. He is restless and, quite unconsciously, may be on the lookout for some one else who will more adequately embody the dream which for the present he has attached to the beloved.

Sometimes the eruption of these suppressed criticisms from the subconscious is most dramatic. The lover may be very fond for months or years or even the major part of a lifetime; the reactions of repulsion may be constantly suppressed and unacknowledged even to himself during all this time, but they have been developing constantly, although subconsciously. Finally, after years of loving devotion, these suppressed repulsions break forth and dominate the personality of the lover. The yoke of a false love is thrown off. His life-long love ends in a moment, completely, because it was never genuine. Although this end of love may seem to himself and to others to come so suddenly, in reality his suppressed criticisms have been undermining his love for years.

On the other hand, there are people who are able to love sentimentally all their lives and are never so devoted to their dreams as at the end of three score years and ten. They are people who have succeeded in destroying their integrity quite completely. They have become wholly unable to distinguish between dream and reality. They have succeeded in dissociating themselves into a sort of double personality, one constantly suppressed, the other dominant and active. They are arch-hypocrites, beyond all hope of salvation. They have succeeded in deceiving themselves so completely that they can never be made to suspect their own self-deceit.

So much for sentimentality. Let us now turn to genuine love, the sort of love which is the essence of Christianity. This love, we have said, is a form of cognition. It is contemplative. It avails itself of all that science can reveal concerning the object of devotion, but it discerns in that object all that which science ignores. That which is useless from the practical and theoretical standpoint is not useless from the standpoint of love. All that practical and theoretical thinking can bring to light is seized upon by perfect love, because it reveals the true nature of the beloved so far as theory and practice can reveal. But love will not be confined to such narrow limits.

"Thou shalt love thy neighbour as thyself." But who is this neighbor whom we should love as ourselves? Is he the foe of society, murderer, burglar, anarchist, capitalist, or whoever may be our pet type of enemy? Is he Dago, German, Wop, or Jap? Is he a social parasite, lolling, lewd and lazy, who has exploited the needs of his country and the poor of his land, a lily-handed epicure who is growing fat on the blood and toil and sweat of those who are ground beneath the wheels of war and industry? Is he that one who holds a grudge against us because he thinks we have wronged him deeply, and watches us with malicious intent, waiting for the moment when we lie at his mercy that he may kick us while we are down? Or is he the man whom we ourselves watch with such malicious eyes because of a wrong which we believe he had done to us?

Jesus has described this neighbour for us in parables and other teachings in such a way that we can recognize him wherever we meet him. The neighbor whom we are to love is anyone who embodies, however unconsciously to himself, however overlaid with prejudice and passion, however perverted and misconstrued, that uniquely human impulse toward more abundant life which can only find fulfillment in the great community of love. Whoever is driven by a divine discontent that will not let him rest with any earthly good save only complete self-surrender to unbounded love, is my neighbor. Whoever is made for love and is restless till he gives up all for love, is my neighbor. Whoever is driven by the thrust of an impulse for fuller life into murder and robbery and cruelty and lust and vanity and self-aggrandizement, exploitation of the weak, and violence toward the strong, he is my neighbor, because that impulse for a fuller life can find the scope and power it seeks only in the great community of love. This purpose of God in every man, this nature which is made for love and for nothing else, which blindly gropes and inarticulately strives, knowing not that it is only love which can give it the amplitude it seeks, this is what is holy in every man. This is the divine spark in him. That he should be so made that he can never find peace until he goes and sells all that he has for love, this is what makes him lovable. And this is what makes him sinful. For sin is nothing else than missing the mark by failing to achieve that richest abundance of life which his nature craves.

What we love in our neighbor, when in the depths of his wickedness, is his need of love and his latent capacity to love. For his sin lies just in this that he has missed the way of love, and so gropes and stumbles and cries in the night, and flings about in desperation until he finds that way, or else is lost forever.

There is nothing in the world worth doing save to show men the way of love. A day's work or cup of cold water is not worth the doing or the giving if it be not given in the name of Love. To foster the flickering, faltering glow of love which is in every man including ourselves, to feed it with truth and fan it with devotion, that is the only work that is humanly worth the doing. And if ever that spark springs into flame and flares aloft it will set the world ablaze and all the gates of hell cannot prevail against it.

But if love be the whole end of living, the Christian way of life, and the way which Jesus taught and lived, let us look a little more closely into its nature.

To love a person means to react to him not only as he appears at the present moment, but as he was in the past and as he will be in the future and as he has shown himself to be in many different times and places. We love the total individuality, and that can reveal itself only throughout wide reaches of time and under many different circumstances. When we love, the past is with us, the future is with us, the events of many different times and places are with us. All this, of course, is not clearly in our conscious minds, although some of it may linger at the fringes of consciousness. But it is all with us in the sense that our attitude toward the beloved is an attitude of recognition in him of characteristics which have been or will be displayed in various times and places, but which are not all manifest at the present moment. The confining barriers of immediate time and space flee away. A whole life becomes the immediate object of our concern. The present, for love, is not merely the passing second, or the passing quarter-second. It is not the passing hour nor day. The present, for love, is the total lifetime of the beloved, or as much of that lifetime as the lover is able to comprehend. Perhaps only love at its best, which we may call perfect love, contemplates the total individuality as he has unfolded himself in the past and will in the future. But love, according to the measure of its excellence lifts us out of the thronging change of circumstances into the wide still view of a total life.

When the loved one is at the uttermost depths of shame, the lover sees him just as clearly and vividly as he was when he was most honorable, or as he will be when he attains the highest honor that is in store for him. For perfect love this honor is just as real and just as present as the shame. When the criminal comes home to his mother, if she has this high degree of love, she knows him not only as a criminal, but also as the infant which she held in her lap years ago. And this charm of his infancy is just as real and present with her now as is the shame of his crime. And more than that, the character which may develop out of this criminal in the future, is for her a part of the total fact of his individuality. Love comprehends wide reaches of time, because love knows the individual in that fullness of his nature which only wide reaches of time can reveal. Love discerns personality; that is, it finds the past operative in the present to shape the future.

Because love can see in the worst that the loved one displays, the presence also of the best, love has great power to arouse this latent best and make it burgeon forth. There are two ways of attacking the evil in a person. One is the direct attack. It may consist of punishment or reprimand. This is not love's way. It may consist of drilling, instruction, admonition, or some other corrective measure. The other way of dealing with the evil may be called the indirect. This is love's way. It consists in turning away from the evil, as it were, and appealing to that latent nobility and excellence which is just as truly a part of the individual as the evil, howsoever dormant. It is only love that can make this appeal because it is only love that can see with sufficient clearness and vividness the living presence of this latent nobility. One can appeal to the general possibilities for good which, presumptively, are resident in all mankind. In so far as all people are alike, having the same potentialities, this sort of appeal will do the work. But only with respect to very general and abstract features are all people alike. When it comes to the full rounded concrete good which may reside in every man, each

is unique.   Only personal association can reveal
this unique, concrete, total good which lives in each
individual; and only in the personal group can one
bring it forth most completely by his appeal of love.
This appeal is generally a mute appeal.   It consists
in the lover letting the loved one know that he sees
this good, and clings to him for it, and yearns over
him for it.   The wordless, devoted constancy of
love is itself the appeal.

As the lover sees the best in the loved one when
he is at the worst, so also the lover sees the worst
in him when he is at the best.   This is not true of
sentimental love, and perhaps most human love has
in it a touch of sentimentality.   In so far as love is
sentimental it is blind to the faults and, for that
matter, even to the virtues, if they happen not to
be pleasing to the lover.   But love approaches perfection only in so far as it is clear and complete
in vision, discerning the weakness as well as the
strength, the faults as well as the virtues.   Hence
in perfect love one will see the latent hidden evil,
even when the loved one is displaying the noblest
that is in him.   Here again, of course, it is not necessarily a matter of conscious recognition.   It is
rather a mental attitude which is adapted to an
individual in whom is evil as well as good.

Now when we speak of a fault or an evil in an
individual we are using language that is not altogether scientific.   Any characteristic is faulty or
evil only when judged by some particular standard.
But what is condemned by one standard may be
highly prized by another.   By what standard does

the lover adjudge these traits or types of behavior to be faults and evils? Are not all our standards too narrow and abstract to take in all the diversity and significance of concrete individuals? And there is perhaps almost as much error as truth in our condemnation and praise, especially when it comes to such a complex thing as human personality. Is it not the prerogative of love to divest itself of those rigid narrow standards and appreciate the fullness of the individual without pruning him down to these artificial and wholly inadequate standards? Our answer must certainly be in the affirmative. Love must not and cannot apply to the loved one those conventional abstractions by which we ordinarily distinguish right and wrong. Nevertheless there is a standard by which love is able to distinguish the good and the bad in the loved one, howbeit a very different standard than that of conventional morals.

Let us endeavor to define love's standard. It is the harmonious and unified individuality of the loved one. But we humans are rarely if ever completely unified and harmonious within ourselves. We are almost always inwardly at war with ourselves, to some degree. What we would not, that we do; and what we would, that we do not. Now it is this inner conflict in the loved one which the lover seeks to correct. This is the fault in him, the evil, which love must strive to overcome by transforming the warring propensities into the harmony of a unified character. When the character we display at one time is irreconcilable with the character we display at another, love is grieved. The lover seeks for us the peace and power of a unified personality, that inner harmony which constitutes what may be called the true self. The moral standard of love cannot dispense with the help of scientific method.

It is impossible that two conflicting characters in one person can both be good, because if one is good the other must be bad since it wars against the first. Both can be good only when they sustain one another or, at least, do not conflict. On the other hand, both may be bad in the sense that both must undergo modification before any harmony is possible. If the lover values one, he must condemn the other, because it destroys the one. Or, if he values both, he must condemn the conflict which breaks down both. Or, if he values the conflict, he must condemn the rest of the personality which suffers from the conflict and seeks to escape from it.

Here, then, is love's standard. It is the harmony and the complete development of the total individuality. In so far as the individual is at war with himself, and hence is destroying himself, the lover must endeavor to remove the inner conflict and save the total individual whom he loves. The fault or evil, which the lover sees in the loved one, is any trait which prevents this inner harmony or causes inner conflict. He who did not seek to correct in the beloved that which was doing him to death would be a false lover.

Now this is the suffering of love—to love one who is destroying himself. This is the suffering of atonement.   It is this suffering of the lover which constitutes the mute appeal of love.   Herein lies the transforming power of love.   It is this which drives out inner conflict, transforms warring propensities, and reshapes them into harmony.   This it does when it causes the beloved to turn against that in himself which causes such suffering in the lover. When the beloved sees that the lover has so identified his good with the beloved that the latter, in pursuing his course, is destroying in himself that which is most precious to the lover, he may be moved to abandon such a course as he could not be induced to do in any other way.   Above all, such love given to him may show him the way of love as nothing else could do; and the way of love, we have seen, is the only way of salvation and fulfillment to human nature.

Let us see a little more clearly what is involved in that transformation of the individual which may be brought about through suffering love.

A man who has committed a murder once, may do so again.   That murder has ploughed channels through his nervous system and mental habits. Furthermore, it has revealed in him a propensity which may lie latent throughout a lifetime, but which requires only the proper environmental conditions to stimulate it into action.   And all the tenderness he may display after the event does not eradicate from his character that murderous trait, just so long as it is latent within him in the sense that it could be evoked again if the proper stimulus should be applied.   But if he is transformed in that particular, it means that the impulses of his nature are so reshaped and reorganized that no murderous impulse could again be aroused by any sort of situation whatsoever.

If such a transformation should occur, then the past murder, together with the latent tendency to murder, ceases to be a part of one's self. It is not the past of one's new, transformed self. It is an event in one's physical past, but not of his personal past. It is no longer a proper subject for grief, remorse, repentance or regret. One grieves over his past only because his past shows what he truly is. It shows what is latent in him and can be elicited from him under proper stimulus. But when the physical past is no longer a part of one's personal past, when it no longer shows what the latent impulses of one's nature may be, because of the transformation of one's nature, then it is foolish to grieve about it. One's sin has been "washed away"; he is no longer guilty. The past event is no more a part of his personal past than the murder of Julius Cæsar, and should arouse in him no more sense of guilt or remorse than does that bit of Roman history.

We see here the Christian doctrine of cleansing from all sin through the atoning sacrifice of divine love. It delivers from sin in so far as it causes one to turn in revulsion against the evil in his own nature because of the suffering which that evil causes one who shares it with him through that community of interest which love involves.

We have said that the lover reacts to the past and

future of the beloved as much as to the present; that, for the lover, the past and future are the unveiling of the true and complete nature of the loved one. But when the nature has been so transformed that certain portions of the past no longer display what is latent in the nature of the loved one, then those portions of the past no longer enter into that total event constituting the individuality contemplated by the lover when he loves. These portions of the past belong to the total event of nature, to be sure, but not to that personal event which is the past operating in the present to shape the future and which is the beloved personality. So it is that the atonement of love sifts out that totality of past, present and future, which is loved, purging it of all that mars its beauty or fills with pain, leaving it altogether fair.

There is a strange transmuting power in love by which physical pain and mental distress become precious to the lover and by him gladly welcomed. This must not be confused with that morbid state of mind in which people sometimes perversely seek pain. The morbid may seek pain because he thinks he thereby glorifies himself, or compensates for some wrong, or attains some excellence for himself; or some psychic knot may have twisted some impulse in such a way that it automatically plunges into pain somewhat as the moth plunges into the flame. Certain perverts seem to delight in pain because of a peculiar psychic bent that has been given to the impulses. But none of these are instances of love.

Love's transmutation of pain into blessedness

appears only when the pain is an experience by which I enter more deeply into the heart and mind of him whom I love. When pain is the gateway into his soul, it is joy. When all my experience, including physical pain and mental distress, becomes a form of community, it is that in which I delight. When I experience nature not merely as physical fact, but as physical fact which is experienced by another whom I love, then nature becomes glorified to me. If this hill is not only a hill, but the hill on which my beloved has trod, then I, in treading upon it likewise, am experiencing what my beloved experienced. I am sharing his life. If the mental anguish which assails me is that which my beloved has experienced, then I am entering into the deeper recesses of his mind and heart when I also experience it. Prior to this experience there were regions of his life I could not enter. There were things I could not understand. He opened the door to me, but I could not come all the way into his heart. I did not know the way. I had not the experience which would enable me to know and see and feel what he knew and saw and felt. But now I can enter in, even into the innermost sanctuary of his soul, because I too have experienced. In this way the anguish, while not ceasing to be anguish, becomes also a joy.

This is not to be confused with the adage: "Misery loves company." It is true that misery may be lightened by knowing that others are suffering also. But this has nothing necessarily to do with love, as shown by the fact that these persons need not be peculiar objects of my affection in order to lighten my misery by their company. It is balm to one's wounded pride to know that one is not the only person who must suffer. But there is a vast difference between this, and the joy of community in which all experience, whether originally pleasant or unpleasant, is transmuted through love.

Love passes its fingers over all the objects of the world and transfigures them. As king Midas by his touch could change all things into gold, so love by its magic can transmute all experiences into something blessed.

> Love is like the wind that passes
> Fingers through the leaves and grasses.

There is a grace on all things that was not there before, because in all things I am experiencing the heart and mind of him whom I love. And to share his mind is my supreme joy. Surely goodness and mercy shall follow one all the days of his life if he loves any one in such fashion that all experience becomes a form of deepening community with him. If pain and anguish are among the most engrossing experiences, and if community in the most engrossing experiences is the deepest and sweetest community, then it is out of the deepest pain and anguish that love rises most triumphantly and shows most gloriously its power to overcome the world with all its ills. Out of deepest suffering may blossom beauty and joy, if there be love.

Some doctrines concerning the Cross are not without meaning in the light of this truth. Perhaps

the greatest message of Easter, and the deepest meaning of the doctrine of the Resurrection, is just this
triumph of love over suffering and death.   At any
rate it becomes apparent how all things might work
together for good unto one who had a Lover of
such a nature that all experiences became a form of
community with Him.   There is no human love
which can transmute all experience in this way.
But great love does it more, and little love does it
less; and we can assume that Perfect Love would
do it completely.

May not this transmuting power of love, and
atonement through suffering love, be the true solution of the problem of evil and sin in a world ruled
by an almighty good God.   To enter into the love
of God is not to abolish evil but to transmute it
and triumph over it.   And sin can be conquered
only through suffering love.   Is love, after all, the
only way to triumph over sin and evil as, perhaps,
God does eternally?   And we can enter with Him
at any time into His triumph not by might nor by
power but simply by love.

Thus love becomes the way of salvation and the
meaning of religion.   To be even one of the very
littlest ones in the Kingdom of Heaven is to be un-
conquerably blessed, for love transmutes all pain
into sweetness and death into life and shame into
good.   To enter into the Christian Way of life is
"to fall in love with the universe" and to find that
God Himself is love.   To know God is to know
love.   In the Christian Way of life there may well
be pain and death and shame, but they are trans-

muted. Pride, that suffers the gnawing tooth of envy, and wounded vanity, cannot live with love. The very least of these in the Kingdom of Heaven has overcome the world with all its ills. The mighty man must break at last but these little ones are dauntless eternally. One can meet them in obscure corners of the earth and it is good to look into their eyes.

Professor Simkhovitch has sketched the historical conditions[1] under which Jesus brought into the world the Christian way of love. The chief importance of this sketch for our present purposes is that it shows, we believe, the only conditions under which men can find this way of love. For it is an exceedingly difficult way to find although it is the only way in which human life can be delivered from all its physical and spiritual ills of pain, vanity, disappointment, fear, futility, bitterness; and the only way in which it can enter into peace, power, fearlessness, and quenchless joy. For even the suffering of atonement has its joy; through the vast community of life which it offers, it gives that vision and power, that maximum abundance of life, which is the supreme satisfaction of human nature. But to find the conditions under which this most excellent way of life (which is as "rare as it is difficult") may be found and entered by men, let us turn to this sketch by Professor Simkhovitch.

The Jewish people for centuries had been developing a supreme hope and aspiration. It is not possible to experience to the utmost the distinctively

---

1 Simkhovitch, *Toward the Understanding of Jesus*, Vol. V.

human ills of life, such as sense of defeat, bitterness, disappointment, futility, envy and other such forms of anguish, except as one is filled and moved by such hope and aspiration. The Jews had it. They were the chosen people of God. They had a supreme mission to fulfill. They had a part to play in the life of all mankind which was to be of the greatest importance. Of this they were sure. And they knew that this great history-making work was bound up with their religion, their distinctive culture, their way of life. But while filled with this great hope, they found themselves baffled and beaten. They were among the most mean and insignificant of peoples and never was this sense of their littleness and insignificance more clearly brought home to them than in these days before, and during, the life of Jesus when they were crushed under the mighty power of Rome.

Finding themselves beaten and all their life and history about to come to nought, different ways of escape were sought by which they could save their hope and heritage, or at least escape complete destruction.

Some gave up in despair and said: What's the use? Our hopes are dead. Our purposes are come to nought. Let us settle down and make the most of what is left. Let us eat, drink and be merry for tomorrow we die. Let us at any rate keep a whole skin and be as comfortable as we can. Let us seek the goods of this world and no more cherish the hope of fulfilling any lofty purpose in the world.

There were others who sought a way out by

open and desperate rebellion. Rome was all powerful. No earthly might could resist her. But these
were the Zealots; they were ready to sacrifice everything in a last desperate fight—home, wife, child,
limb, life—violence to the last, and then death.
But perhaps God would intervene at last, before all
was lost, if they endured to the end.

Still again were the Sadducees, who took the opposite course. They would compromise and adapt
themselves as best they could to the situation. They
were the worldly wise. They would keep what
they could of their culture and aspiration, but
would also take over what seemed of value in the
Hellenic civilization. They would adapt their Jewish hopes and aspirations to the world as they
found it and win the prosperity and success which
conditions offered.

Jesus seems to have considered all three of these
possible solutions of the problem. Such is the interpretation Professor Simkhovitch gives to the
temptation in the wilderness. He could turn stones
into bread. His people could give up their high
historic hopes and live as comfortably as possible
under the circumstances. He might minister to their
bodily comforts and let the dream of an unfulfilled
mission pass over. Or he might lead them with the
Zealots in desperate revolt against Rome. Yes, he
might dash himself down from the pinnacle of the
temple. Such an undertaking was hopeless from
the earthly standpoint, but God would intervene
and the angels would bear him up lest he dash his
foot against a stone. Or finally, he might appro-

priate the power and culture of the Hellenic civilization of his time, keeping what could be merged
with it of the Jewish culture. Accepting thus the
kingdom and glory of the world he might bow
down and worship Satan. But this course he also
rejected.

The course he finally adopted was closer to that
of the Pharisees than to any other. The Pharisees
would not resist the political power of Rome, but
under this yoke they would cherish all their customs and their religion with the utmost tenacity.
Rome might wield her earthly power, but they
would hold fast their spiritual heritage, their hopes,
their culture, their ideals. If they kept these pure
and undefiled, so they believed, the Messiah would
come and deliver them from the foreign power.

But their efforts were doomed to failure and this
Jesus plainly saw. Under the Hellenizing influence
of the Roman power their traditions were being
gradually but inevitably contaminated. At best
they were being reduced to little else than the external forms and ceremonies. Like an ineradicable
acid eating into their culture, the subtle influences
of Hellenism were permeating their lives. They
could keep the external forms, but the spirit would
surely go unless they took the offensive. They
could not merely stand fast and hold tight. Either
they must permeate the world with their spirit or
be permeated by it. They did not resist with violence, neither did they compromise, neither did they
give up in despair their hope and heritage and content themselves with mere animal existence. In all

this they went with Jesus and he with them.   But
there they stopped and he went on.   While they
did not resist their enemies save only hold fast to
their traditions, they had bitterness in their hearts
toward the conqueror.   They were crushed and
baffled and beaten because they did not love their
enemies.   They could not conquer Rome because
they could not love enough.   They could struggle
against the alien influence, but they could not reach
out and transform that alien influence, for only
love could do that.

So it was, in the fullness of time, that to a baffled
and beaten, a desperate and despairing people, and
himself one of them, Jesus brought his way of love.
There is no way out, he said, but this.   It is love.
Love your enemies.   Do good unto them that despitefully use you.   Who were their enemies?   Who
were despitefully using them?   No Jew could mis-
understand the reference.   The enemies of Israel
and of every Israelite needed no further specifica-
tion.   They must turn the other cheek.   They must
love their neighbor even as themselves.   And who
was the neighbor?   Not merely a fellow Jew, but
an alien also—a Samaritan—could be neighbor to
a Jew.   In the world ye have tribulation, but be
of good cheer, for I have overcome the world.   And
how overcome the world?   By love.   There is no
other way to overcome it.   By love shall ye conquer, and by love alone.

We have looked into the historical circumstances
in which Jesus brought to men the way of love in
order to point out what we believe to be the only

conditions under which men can be led into this way.

First of all there must be stirred up in men high hopes and aspiration. They must be induced to seek that more abundant life, the seeking of which makes them more than beasts. They must feel the urge of that discontent that drives them on to something further. But this is only the first step.

As long as men are successful and prosperous in their efforts to achieve what seems to them a more abundant life, they will not turn to love, for that is not the natural way. Of course all men are affectionate to some degree and toward some people. But they do not naturally seek to realize the community of love. On the contrary, as long as the world favors them their ambitions will be competitive. They will measure their success and power, their fame and fortune by how much it overtops that of others. Therefore they must seek their good at the expense of others. They must exercise a power which is not shared by others. They must possess goods in such way as to exclude ownership by others. They must have an honor which makes others appear less honorable in comparison. This is the natural way of seeking the more abundant life as long as success attends their efforts.

Hence a third thing is necessary before they can find the way of love. They must suffer defeat. They must feel themselves baffled and beaten. They will continue to pursue the natural way as long as that way is open. They will not cast about for another way until that way is blocked. Hence the necessity for failure, or disaster of some kind, such as the Jews experienced. Such as Jesus himself experienced and by which he opened up the more excellent way.

But still a fourth thing is necessary. When men find themselves overcome and overwhelmed, they must not give up in despair. Neither must they turn to bitterness and envy, tenaciously holding to their aspirations, but filled with disappointment and jealousy, as were the Pharisees. Neither must they compromise, as did the Sadducees. All three of these ways are the common courses followed by men in the hour of defeat and failure. Neither must they turn desperately to violence as did the Zealots. But they must seek persistently to bring to fulfillment their high endeavor until they find the only way in which it can be fulfilled, and that is the way of love.

When we consider all the obstacles that must be overcome, and all the devious ways that lead astray from that of love, and how difficult is the process which leads to it, a flood of light is thrown on the saying of Jesus: Wide is the gate and broad is the way, that leadeth to destruction, and many are they that enter in thereby. For narrow is the gate, and straitened the way, that leadeth unto life, and few are they that find it. This straight and narrow way is the way of love and few indeed are they who find it. But every other way leads to destruction.

Now can men be brought into this way of love? They will not enter it or even note it until they find the natural way of competitive goods to be blocked and impossible. Can nothing then be done until the hour of defeat and trouble? No, much can be done. Even in the hour of defeat men will not ordinarily find the way of love, but rather will turn to despair, or bitterness and envy, or violence, unless they have been taught of this way of love. But having been instructed concerning it, above all, having had it exemplified to them by the lives of others, it will come to their thoughts in the time of perplexity and they will turn and seek it out. Jesus himself had the teachings of Isaiah concerning the suffering servant, for Isaiah also lived and taught in an age of defeat and despair. We have the lives and teachings of many great lovers. And there is no great ministry to men save to augment this teaching and example of love in order that greater numbers will turn in the time of bewilderment to the only way that is open to the human race.

When one looks out upon the human race, the way it has come and the way it must go, and sees that tiny gate so obscure that one must search to find it, and so lowly that one must stoop to enter it, and yet the only way to life, the only escape from ruin for mankind, one is sobered. One cannot hope that there will be continuous days of easy power and prosperity, for in such times men miss the way of love, and this automatically brings on destruction and the end of such comfortable periods. Civilizations will be transitory until men in large numbers go this way of love; and then that which shall arise will be so different from what we today call civilization as to require another name.

PART II
WHY SCIENCE NEEDS RELIGION

CHAPTER V

THE TWO SIDES OF LIFE

Scientific method and religious experience are the most extreme expressions of a duality that runs all through human living. All the way from the lowest biological level of the human up through the social to the highest spiritual achievements, we find the opposition of these two contrasting demands which human life makes. The contrast may be called roughly that of efficiency versus appreciation; or that of adaptation versus creativity.

At the biological level we find fixated habits, due partly to the innate disposition and partly to discipline. These stereotyped forms of behavior are generally adapted to certain features of the environment. They generally operate with the minimum variation and with a high degree of coördination and efficiency. But over against these habits or instincts, or combination of habit and instinct, human nature is endowed with a great capacity for the free play of impulses, impulses unconstrained by any established system of adaptive behavior. These free impulses represent initiative, change, adaptation to new and more diversified features of the environment, and appreciation of richer and fuller fields of experience.

Passing on to the level of society we find the same contrast between opposing sides of life. On the one side we find the coördination of impersonal association, where each individual and each unit of behavior fits like a cog in the great social mechanism; where each turns the other man's mill but without mutual sympathy or understanding or community of purpose, without appreciation of one another's experience nor stimulating interchange of thought nor integration of ideas. Over against this impersonal type of social organization is the personal in which there is mutual recognition of individuality, where there is provision for free and creative play of impulse on the part of each, and stimulus to such creativity and richness of experience because of mutual recognition and appreciation of individuality.

Going on to the level of the intellectual and spiritual history of man we find still again this counterbalance of opposing interests. On the one hand is the effort to reduce the world to mechanistic and materialistic terms, for it is only as we reduce a process to those terms that we can calculate and predict results with accuracy, control with certainty and think through in clear and definite terms. No systems of thought have been so clear and definite as the materialistic. Materialism is the assumption that the universe is nothing else than those mutually exclusive elements that can be added, subtracted and generally treated as "equivalent." But over against materialism and mechanism has stood the demand for spirit and purpose. And this latter has been no less insistent, undying and universal.

Let us be very clear that materialism and mechanism, as well as spirit and purpose, are postulates or assumptions made in the interest of certain great aspirations of the human mind, the aspiration, namely, to predict, control and clearly define everything in the universe. We can never do justice to materialism and mechanism until we see that it is the expression of a great hope and a great effort, just as truly as the spiritual and teleological interpretations of the world. The materialistic and the spiritualistic demands both represent at the high intellectual and spiritual level the two contrasting ways in which human life labors to uplift and magnify itself. One is just as brave and adventurous as the other. Both are ventures of faith. Great gain and loss is at stake on both sides. Under the banner of mechanism and materialism men have gone forth in the high faith that they could predict and control the changes of nature and could formulate those imaginative, intellectual experiments by which science thinks its way throughout the universe. This has been a great enterprise and it would be sad indeed if men were forced to give it up and pronounce the faith unfounded. But just as heroic is the enterprise on the other side and just as much is at stake. Under the banner of spirit and purpose men have gone forth seeking in the universe that which is beautiful, that which can be loved and that which can be worshipped. Is beauty merely something that we attribute to nature, or is it resident there just as truly as the molecules or any sense object? And is there in the universe, as basic and far reaching as the elements of matter or energy, that which can be loved and adored?

Nothing can be more pitiful than to see those marching under the one banner ridiculing those who uphold the other as though those others were vainly following an illusion, quixotic in their folly. As much faith is involved in the one as in the other. As great a good, perhaps, is at stake in the one venture as in the other. We would even go farther and say that if the one cause is hopeless so is the other, for men cannot long continue either enterprise without the help of the other.

Hence arises the supreme problem: How to adjust these two views, these two faiths, these two enterprises, in such fashion that they shall not necessitate an either-or, but can be taken as both-and. They must be rendered mutually inclusive, no longer mutually exclusive. Can materialism be transformed into that concept of nature which provides for accurate prediction and control but which does not exclude the quest for personality, beauty and love in the universe at large? And the concepts of personality, purpose and beauty be so shaped as not to exclude from nature the possibility of that which science requires to find there in order to do its work? The required adjustment between the two views of nature has been made frequently enough, to be sure, and many people think the problem is solved. But it has never been solved with complete satisfaction to both sides and probably will not be solved finally for a long time. The most we can hope is to move toward more adequate solutions.

Here, then, are the two poles of living. On the one hand efficiency and adaptation, on the other appreciation and creativity. The development of disciplined habits stands over against the fostering of free impulse. The efficient coördination of impersonal association contrasts with the creative self-expression of the personal group. Above all, that most rigorous constraint of impulse which we find in scientific method is opposed by that most complete expression of individuality which we find in love and religion. We have found a common meeting place for these two sides of life in experience. We have found that they differ, in terms of experience, in that the latter is concerned with the most concrete fullness of experience while the former deals only with certain selected and refined data. Having stated the contrast in general terms let us go a little more carefully into the three levels of this contrast.

THE BIOLOGY OF THE TWO DEMANDS

In original human nature there are certain instincts or habit-forming propensities which are innate. Also innate are loosely organized or quite unorganized impulses. Perhaps it would be more accurate to say that while the first crop of impulses become established, in part, as habits, a new crop of impulses not yet so established is being ever anew quickened to activity. However we express it these two sides of life have their roots in the biological nature of man. New unorganized impulses do break forth over and above the established system of habits. It is these free impulses which make us responsive to the wealth of experience which lies beyond those special features which stimulate the instincts and established habits. In other words, these free impulses are what open our minds to the wealth of experience, just as the habits close our minds to all save a few features.

How to adjust to one another an efficient system of habits, on the one hand, and the free play of impulses, on the other, is one of the serious problems of human living. [1] John Dewey has devoted a book to the study of how to provide for free play of impulse in such a way that the established system of habits can be constantly modified and progressively reorganized into an ever more complex system through assimilating such free impulses. Human life requires a pliable and progressive system of habits; but this is possible only when there are many new impulses springing forth to suggest new ways and to lead on to an ever richer life. The problem is twofold: (1) How to foster the free play of impulse while at the same time maintaining a disciplined system of efficient habits; (2) how to keep the two—habits and impulses—in fruitful and intimate connection with one another, so that impulse will prevent habit from becoming unchangeably fixated and habit will keep impulse focused on the actual problems of living. The second of these two is quite as important as the

---

[1] *Human Nature and Social Conduct.*

first and yet its importance is not so widely appreciated.

This problem of how to promote and adjust disciplined habit and free impulse is most commonly presented as the problem of how to promote and adjust work and play. Work is disciplined habit; play, or recreation, or leisure, is free impulse. There is wide recognition of the fact that human nature cannot dispense with either. Recent thought is most marked by increasing recognition and emphasis on the value of play. Almost every one now sees the importance of developing capacity for both in every individual and providing opportunity for exercise of both.

But everyone does not see that this is only half the problem. It is not enough that each individual should give a certain portion of his time to the exercise of disciplined and efficient habits in some productive work, while another portion of his time is given over to free impulse. To keep the two separated, neither influencing the other, is disastrous to both and to life generally. The chief value of free impulse is that it may make possible some larger measure of spontaneity, creativity, interest and joy in productive work. And the chief value of disciplined habit, over and above the production of necessary utilities for consumption, is that it will so direct free impulse that the latter will produce materials and conditions for its own progressive amplification. Free impulse wholly bereft of the guidance of disciplined habit, destroys itself. It gets nowhere; it becomes vapid and inane, like blowing bubbles that flicker out as soon as produced. Furthermore, it soon begins to consume more than it produces, unless checked. But worst of all, free impulse without guidance of disciplined habit falls into conflict, one impulse destroying the good of another. Free impulse becomes nothing but vanity and vexation of spirit when wholly separated from disciplined habit; but disciplined habit becomes a horror of soul-killing routine when wholly separated from free impulse.

The adjustment of habit and impulse most frequently made, because of the difficulty of the proper adjustment, is to separate the two. "Work while you work and play while you play," is the formula for this wholly unsatisfactory arrangement. L. P. Jacks, in recent issues of the "Hibbert Journal" has been showing the futility and disastrous consequences of this division. He speaks of disciplined habit or work as "belly filling," and free impulse or recreation as "soul saving." The modern world, especially since the introduction of machine industry, has been turning more and more to this water tight separation between work and play. Shorten the hours, increase the pay, devise a system where required production can be provided with minimum labor, and then, however dehumanizing the work may be, we can preserve our humanity by proper use of leisure, so the slogan runs. Give over the work hours, the shop, the industrial region, to the stupifying fulfillment of disciplined habit; but provide time and place and ample opportunity for play, where we can recompense ourselves for the hours of misery, or, what is much worse, the hours of contented stupor, by releasing free impulse in the form of pleasure.   At best this free impulse in hours of leisure may take the form of art and worship and pleasant social intercourse; at the worst, the gross and destructive forms.   It is hoped that men and women living such divided lives can be "educated for leisure" so that they can find recreation in wholesome or at least harmless forms.

L. P. Jacks and many others have shown the impossibility of any such adjustment by separation. The bellies never have been and never will be filled by any such method, says Jacks, or can the souls be saved.   Work in which there is none of the spontaneity and creativity of free impulse becomes uninteresting or positively miserable.   But where work is not interesting it cannot be highly efficient. The worker will certainly produce less and less compared to his possible maximum.   At the same time free impulse, altogether separated from work, inevitably becomes more and more wasteful, consumptive of ever more and more.   So, with this separation, work must become less productive while play becomes ever extravagant increasingly.   The amount of utilities which an ordinary sophisticated man of today requires in order to "have a good time" is mounting up to astounding proportions.

But that is not all.   When we turn away from work to hours of recreation for our enjoyment, for the interest and zest of life, the work will become more and more disagreeable to us.   Precisely because our minds and hearts are elsewhere, because our dreams and hopes are elsewhere, because we take it for granted that work has no value for us except as a means to an end, because we do not try to put any free impulse and creativity into it, it must degenerate to lower and lower levels of spiritual degradation. This gravitation of work to ever lower depths of hell is the inevitable outcome of any such arrangement.

But the indictment still continues. As work becomes more stupifying and degrading, it destroys our capacity for enjoying our play. Some may think that the more disagreeable the work, the more enjoyable must be the play in contrast, but such is not the case. To enjoy one's self in hours of leisure is a very fine art and requires a finely equipped mind and body. But work of the sort we have been describing makes impossible any such finely equipped mind and body. Pleasure, as we see today, becomes more vain, superficial and unsatisfying as work becomes more destructive of spiritual capacity. We may "educate for leisure" with might and main, but we cannot offset the evil conditions of life by any purely academic process unless the academic process is directed to the correcting of those evils.

Finally, work which has not interest and enthusiasm in it will become more and more the haunt of profiteering ghouls. If the mind and heart of the workers are not in the work, if they are not interested in its management, its improvement, its reorganization, they certainly will not protect it from evil management and from the exploitation of those who will use it altogether for their own selfish ends.

Now what has all this to do with science and religion? The connection may not be very apparent but is in fact very intimate and vital. The promotion and adjustment of the two sides of life we have been considering is the same identical problem as that of how to promote and adjust science and religion.

Science, both pure and applied, is the development and discipline of efficient habits, habits so shaped as to provide most effectively for biological adaptation to physical environment, for social coördination in industrial production as well as elsewhere, and for the control of natural processes and for accurately calculating results. Science is simply the extension of the old common-sense method of trial and error by which habits were disciplined and shaped to the ends of most efficient adaptation to environment. The method of trial and error tests the impulses when once they are awakened, but it does not awaken them. That the total situation must do, if it is done at all. Scientific method is not a method for awakening the free play of impulse. It is true that science cannot advance except as there is this spontaneity, for it is only by these novel impulses that new theories can be presented for scientific treatment, and new ways of doing things suggested for the experimental tests of scientific method. This is to repeat what we have stated all along, that science requires, for its own maintenance and progressive advancement, the support of that which it cannot itself provide. Just as the disciplined habit of work requires to be impregnated, inspired and rejuvenated by free impulses, so must science be supported by them.

But how can free impulses be aroused? How can all the chords of individuality be quickened to utmost spontaneity? How can all the capacity for response, which resides in human nature, over and above the established system of habits, be awakened? Let us be very clear that scientific method is not a method by which this is done. It is true science feeds on such free impulse. It lives by making use of the "happy thought," the illuminating suggestion. But it is not itself devoted to bringing these novel hints to life. Its business is to put these suggestions to good use when once they are provided. Its business is to test by experimental technique, to adapt, to reshape, and fit these free impulses into such form that they will lead on to the solution of the problem. Scientific method can regulate, check, test, guide into profitable channels whatever new impulses may arise. But how can they be quickened into life? We shall see it is in religious experience the utmost capacity for free response is awakened. In religious awareness we become responsive to the fullest mass of concrete experience. It is then that we become exposed to the full impact of the totality that we have experienced.

Now to say that religious experience more than anything else awakens free impulse, and that scientific method feeds on the novel suggestions of free

impulse, does not mean that the free impulses issuing from religious experience as we now find it can
be put to any use by scientific method.   On the
contrary it is precisely because such is not the case
that we have asserted all along that science and
religion are in a bad state of maladjustment.   Just
as the free impulses generated in the sort of recreation that now prevails cannot be turned to account
in industrial production, or work generally, because
work and play are not properly adjusted, so also
in like manner science and religion are maladjusted.
For science and religion are but the more extreme
expressions of this contrast and counterbalance in
life.

But it is conceivable (and at times has actually
occurred) that the spontaneous creativity of play,
and the efficient adaptations of disciplined habit,
have somewhat interfused and have served to guide,
inspire and enrich one another.   Such an adjustment also between science and religion is conceivable, and is more or less remotely approximated in
some cases.   But before any such ideal state of
affairs could occur it would be necessary to readapt
both science and religion.   Religion would need to
become respectful and receptive to the theories and
conclusions of science.   On the other hand science
would need to change some of its basic assumptions
as to the status of matter and the clearly defined
objects of intellectual cognition as over against the
total fact of experience.   This change is under-
way, as we shall try to show in the next chapter.
But even under the most perfect conceivable adjust-

ment between science and religion there would still be many impulses arising in religious experience which science could never turn to any use. This would be partly due to the necessary limitations of science; but it would also be due to the inevitable "wildness" or error involved in all free impulses of human nature, whether aroused in religious experience or elsewhere. Religion needs the constraint, guidance and tests of scientific method just as much as scientific method needs the freedom, spontaneity and creative impulsiveness of religious experience.

But let us now go on to what we have called the social level of this same contrast between the opposing demands of human living. In fact we have already entered on the social level in our discussion. It is impossible to treat human nature on the biological level alone. Human nature is largely a product of society. Both the disciplined habits and the free impulses of which we have been speaking are the products of society. The selective attention of the instincts in the human is very crude and inefficient, and the course of procedure they initiate is ill-adapted even to the biological welfare of the individual. They require much refinement and reshaping, and this is brought about by the language, the customary habits, the institutions, and the conventional attitudes of the society into which the individual is born. But above all, the instincts and free impulses must be shaped to meet the requirements of social coördination in the particular social system that prevails where the individual is.

This is the work of social moulding. Now society may, and generally to some degree does, foster the development of both sides of life in the individual. It may stimulate the play of free impulse and promote the organization of a system of disciplined habits. By means of wit, art, play, worship, by means of the intimate personal associations of the small group who can adapt themselves to the unique individuality of each, society may arouse initiative, originality, and the spontaneous play of impulsive personality. On the other hand, by means of the complex coördinating system of the industrial plant, the state, school, church, and even at times the home, the individual may be moulded to fit like a cog in the social process. He may thus become a bundle of disciplined habits without spontaneity or creativity of any kind. But generally, and perhaps to some degree always, society promotes the development of both sides of life.

By turning to the study of society, then, we mean turning to consideration of these contrasting types of social organization which foster, respectively, these two sides of our nature.

THE SOCIOLOGY OF THE TWO SIDES

The sort of social grouping which fosters free impulse, appreciation and creativity we shall call the personal group. That which promotes disciplined habit and efficient coördination rather than creative integration, we shall call the impersonal. These two types of social organization may be contrasted with one another on four counts.

1. In the personal group the individual responds to what may be called the total personality of the other members. By that we mean that each member adapts himself to the thoughts, sentiments, purposes and needs of the other members. Each regards the other as a unique individual to whom he adjusts himself in a special way. Each understands the other. Any association of friends would represent such a group. Such group life, to reach its maximum, requires two things: First, that the complete individuality of each be evoked; second, that each adapt himself to, and appreciate, the complete individuality of every other.

In the impersonal group, on the other hand, there is no such recognition of individuality nor is individuality evoked. Each may respond to some act on the part of the other just as one would respond to some physical object or mechanical signal; each may respond to some function of the other; or it may be merely the spatial position of the other to which one adapts himself. In any case it is not the unique individual purpose, not the shades of feeling, in the other which one recognizes; it is not the more comprehensive thought of the other that one appreciates. The Bell telephone system is an example of an impersonal group of which we are all members if we use the telephones. The bank in which we have money deposited is one of the impersonal groups, including all the other depositors and borrowers as well as the officials. Our life insurance company is another impersonal group, including all the other policy holders and the officials as well as the investors of the deposited funds.

2. The second point of difference between these groups lies in the degree to which the complete individuality is evoked in each. This has already been mentioned in connection with the first difference mentioned, but is worthy of consideration by itself. In the personal group we not only respond to the complete individuality of the others, but our own total personality, approximately, is stirred in the response. Of course this is a matter of degree, for groups are more or less personal. It is in love and worship that this total response of the individual reaches its maximum. But a distinctive feature of the personal group is that more of the habits, sentiments, and impulses of the individual are simultaneously aroused than is the case in the impersonal. In the association of friends and in the discussion group we undergo that stimulation of all our tendencies, which appears in consciousness either in the form of deep, pervasive emotion or else as mental activity, wide, varied, rapid, and spontaneous.

In the impersonal group, on the other hand, emotion is at the minimum and mental activity, although at times it may be very strenuous, is effortful rather than spontaneous, instrumental rather than creative, exercised as a means rather than as being in itself satisfying. All response in the impersonal group tends to assume the form of the minor units of behavior which operate automatically. Motorists on a congested street corner constitute such a group. Each responds to the signals of the road in an automatic fashion. When the man in front of me holds out his hand, indicating that he intends to turn to the right, I adapt my conduct to his with scarcely any emotion or thought upon the matter. In the bank I sign my name to the check, thus responding to the requirements of the group. But only a very few of the tendencies constituting my personality are involved in the act. In the polite society of the ball room or reception hall my conduct becomes an unemotional, unthinking, mechanical fulfillment of the rules of etiquette, just in so far as the group is an impersonal one. But the moment I meet there an old friend, my response is transformed. My conduct ceases to be a matter of a few reflexes and comes to involve all those tendencies that make up the deeper sentiments and more comprehensive purposes of life.

3. The third point of distinction between these groups lies in the plasticity and rigidity of their respective types of social order. The order of the personal group is highly plastic; that of the impersonal is rigid. The order of the personal group is constantly shaped and reshaped by the changing moods and needs and purposes of its members, because its system of organization is nothing else than the mutual adaptation of its several members to the needs and purposes of each. It adapts itself to every personality that comes into it, just as the personality reciprocally adapts himself to the group. Thus it is in the personal group that one finds max-

imum social freedom, if by that one means to be free with other people. But if one means to be free from others, free from the need of considering their wishes or their personalities, then of course he has it in the impersonal system. But this is more a negative freedom, while the personal group provides a positive freedom.

The order of the impersonal group, we have said, manifests rigidity. Its order is prescribed by rules and regulations. It is not maintained by mutual adaptation of persons directly to one another, but rather by adaptation of each person to the regulations or code that is enforced. This code may be designed and imposed by some dominating individual, or by mutual agreement. It may consist of written constitution and statute law, or it may be the product of accumulating tradition. The greater part is generally the latter. But in any case it constitutes a framework into which each member of the group must fit himself, and in so doing his activities are automatically coördinated with those of other members of the group. The members of an industrial plant, for instance, may act in such a way as to constitute a beautifully coördinated system. This system, however, does not arise out of any mutual consideration for one another on the part of the workers. It arises out of the mechanism of regulations by which their activities are coördinated.

4. The fourth point of contrast is in respect to the permanence or change of the personnel of the two groups, respectively. The personnel of the

personal group is constantly changing. The employees of the industrial plant come and go, but the members of the home continue their relations to one another throughout a lifetime. The political order may seem to be an exception to this, for it may be quite impersonal and yet its personnel is fairly permanent. This, however, is due to the peculiar position of the political order relative to the other groups. It is supervisory over all other groups, and hence, wherever we go we are always either actually or potentially under its control. However, it should be borne in mind that the political order is usually less impersonal than the industrial.

It should be noted that actual social groups are rarely if ever purely personal or purely impersonal. Actual groups are generally a mixture of these two types. But it is possible to call a group personal or impersonal according to the predominance in it of one or the other of these two types.

Also it is apparent that in these two groups we have the social aspect of the contrast we have been considering in human life between habit and impulse. In so far as we react to others according to established and automatic habits, our social relation with them is impersonal. In so far as these habits are modified by new impulses quickened by the uniqueness of the other and of the total situation, our relations are personal. The problem here as always is not merely to keep the two types of association from interfering, but to connect them in such a way as to support one another. We cannot have profitable personal intercourse without a basis and framework of habit which constitutes impersonal relations. Love at its best, where the whole personality is deeply responsive, requires a finely standardized system of habits, disciplined and adapted, yet plastic, suffused and supplemented by free impulse. Lovers like Jesus and John, or Plato and Socrates, or Ruth and Naomi, cannot dispense with the standardized system of traditions and regulations of their day; but they cannot be satisfied with these alone nor accept them without modification. Personal association must draw upon all that the impersonal can supply of routine and method, but add to it the warmth and delicacy and new creations of impulse. And impersonal association must not only draw upon the personal for plasticity and adaptation to novel situations and to the unique requirements of individuals, but connect this with the historic achievements of the race so that the social interaction can be borne up by the labors and wisdom of thousands who have lived, and also provide that it may contribute something to the heritage of thousands yet to live. But this adjustment of the personal and impersonal, making possible such mutual enrichment, is rarely if ever perfect. Often the two hinder and destroy one another.

But our thesis is that these two types of social interaction are indispensable to one another. Each contributes something which is necessary to human life. The personal provides social freedom. The impersonal provides discipline which is requisite in order to stabilize human purpose and bring human endeavor into effective action upon stubborn obstacles. Especially is it needed to bring about a summation of a great number of individual efforts upon a common enterprise.

While the personal group engenders progressive evolution of purpose through the quickening of new impulses, the impersonal provides that mechanism of achievement by which such purposes can be executed and so provide conditions and materials for a still larger purpose.

It is the personal group which causes the individual to enter into response to an ever greater number and diversity of qualities in the world around him. It ushers him into a more concrete world. In the personal group we teach one another to appreciate color and sound and delicate shades of difference and all that massive fullness of fact which the impersonal group, in the interests of efficiency, must ignore. In engenders delicacy of response and depth of emotion. But it is the impersonal that provides the interlocking chain of consequences and that wide field of endeavor by which the efforts of the individual are transmitted and magnified to the utmost. With its wide-reaching systems of coördination and its indestructible institutions, it gives world-wide significance and historic efficacy to the work of the individual.

The fertility of the historic social process, out of which has arisen the arts and sciences, is due to the marriage of these two types of association. When one operates in such a way as to impair or destroy the other, both must suffer and human life is impoverished. Our persistent problem is how to adjust them that each shall sustain and magnify the other, instead of thwarting the other, as is all too frequently the case. Since the industrial revoluion, for instance, we have seen the impersonal driving out the personal. But of late a strong movement has set in to counteract and correct this one-sided development. The danger is that we shall set the two over against one another and to the exclusion of one another.

Here again is the problem of science and religion. The organization of impersonal groups is ultimately a matter of science. Heretofore they have been largely shaped by tradition and common sense; but in the fields of industry and government we see science gradually supplanting tradition and the rough rules of practical experience. The transition is necessarily very gradual but it is on the way. We have the beginnings of a political science. There is a most efficient method of governmental administration and political action which science must devise and which is even now in its beginning. There is a most efficient method of industrial production which science alone can discover and set forth. This achievement of science is also under way. We have our engineers of production, engineers of organization for personnel as well as for machines, materials and finance. One needs only mention, for example, the so-called "Taylor system" to suggest the work of science in this field. Impersonal association plainly belongs to the side of science and is gradually becoming identified with science.

But just as truly personal association belongs to the side of religion, the complete individuality can be evoked only through religious experience. This follows necessarily from our definition of religious experience. There are of course religious traditions and ecclesiastical systems which iron out individuality. But religious experience as such does just the opposite. Nowhere do we find such unique, powerful and fully aroused personalities as in cases of vital religion. Great religious leaders have often stood up against mighty institutions, literatures, and all the machinery of the impersonal order and by sheer force of the personal association of others with themselves have turned the stream of history into another channel. This they have done without making any use of the written word, for it is less personal than the spoken; without using the institutions or any impersonal mechanics of society. The origin of Christianity is, of course, the outstanding example of this. But it was not only the personality of Jesus that towered so high. Humble obscure men, when touched with the dynamic of this great religious experience, were turned into giants of individuality. The total capacity for response, latent in the unique individuality of each, was aroused.

And as religion evokes individuality, thus providing the first requirement of personal association, it also, in its Christian form at least, brings to the highest level that love which is the second essential.

Love can reach its maximum only when God is known as love. It can attain that quantity and quality required for the greatest personal association only when it finds its nourishment and stimulus in the spirit of the universe.

There is no science of friendship and there is no hint that there ever will be, for personal association can never be a matter of standardized rules and regulations. It can never have that uniformity which science requires. Personal behavior can never be wholly predictible and its units never "equivalent." There is certainly a scientific adjustment of impersonal organization which is favorable to love and mutual understanding, as there is another that is inimical. Much work is now being done in the field of industrial organization and elsewhere to provide for personal relations and creative interest. This is auspicious. But the actual personal intercourse and the free play of impulses can never be a matter of system and regulations. Spontaneity, creativity, and mutual appreciation, can never be manufactured by scientific method. The fountain of self-expression and novel impulse generated in free personal association may very well be poured into the hopper of scientific method and be transformed into useful and efficient forms of social behavior. That would be an ideal adjustment.

Religion can be propagated effectively from man to man and group to group only through personal association. Jesus sought above all else to have personal relations, and formed a small group about himself for that purpose. All his teachings con-

cerning social relations glorify the personal group.
The early Christians were intensely personal in
their relations to one another. The Christian
church at its best always seeks to cultivate personal
association. There is nothing more deeply personal than for several individuals to share a common religious experience. Christian love is personal association at its fullest.

It is plain that in personal and impersonal association we have these two sides of life that head up
in science and religion.

Let us now turn to what we have called the
third level of this contrast. One might almost call
this the two metaphysical demands. There is a
history of man's intellectual struggle for clarity
and definiteness of concepts in thinking his world,
in controlling and predicting its processes. There
is also a parallel struggle for recognition and appreciation of spirit and purpose throughout the world.
We want to glance at this twofold endeavor for
scientific precision and spiritual significance.

THE TWO METAPHYSICAL DEMANDS

From time immemorial the process of trial and
error practiced by common sense has been carving
out from the massive world of experience an
orderly and manageable little world of practice and
theory. The work of one generation has been
handed on to the next by means of words and conventional attitudes, formulas, habits, traditions,
and institutions. Finally scientific method enters
in to put on the final touches to this tight little

world of neat and controllable objects, which common sense has so laboriously developed throughout successive generations.

> We are presented in sense experience with something large and vague and chaotic, and we sophisticate and conventionalize until we have made something neat and definite and of manageable proportions stand out from the bewildering background of experience. The process of trial and error is not so simple and straightforward as it appears, and a vast amount of unseen work has to be done before there emerges the world of familiar things which we suppose we live in, but which is a theoretical construction from what is actually presented to our senses. The man of science and his common-sense forerunners brush aside a vast number of perplexing problems when they decide to disregard metaphysics and go straight to work on particular parts of the physical world.[2]

But the man of science and his common-sense forerunners never do get rid of metaphysics and never can. Ritchie immediately adds in the next sentence: "But their concept of the physical world is already saturated with metaphysics, and metaphysics of the most dangerous kind, unconscious metaphysics inherited from our forebears or worked out in extreme youth. Lurking in the background of any description of what we see and touch and hear is some theory and some assumption as to the

---

2 Ritchie, A. D., *Scientific Method*, p. 6.

nature of things. The scientific man's escape from metaphysics is largely illusory."[3]

The Greeks began this work of definition, clarification, and exclusion so far as science builds on common sense. With the close of the Middle Ages science developed the experimental method to a degree the Greeks had never been able to do and by means of it we have been able to define, predict and control with an accuracy and range which the Greeks could never approach. And it would seem that science has only begun its great work of constructing a world to suit its purposes out of the mass of immediate experience. So we find ourselves today in a world more clear, predictable and controllable than ever before. Now this world, with the concepts by which we think it and the attitudes we assume toward it, inevitably implies, as Ritchie says, a metaphysics, whether we consciously recognize it or not.

This scientifically constructed world in which we live, scientifically controlled, defined and predictable, carved out of the chaos and massiveness of experience and made to stand forth like a hard little jewel from the mists and shadows and streaming mysteries of immediate experience, is not an unreal world. If we say it is constructed, if we call it artificial, we do not mean to imply that it is fictitious in the ordinary sense of fiction as being a construction of fancy. This hard and definite little world of science and common sense is just as much fact, as far as it goes, as the total fact from

---

[3] *ibid.*

which it is distinguished by means of concepts specially constructed for this purpose. This world of practice and theory stands in the midst of total fact somewhat as a cubic foot of air stands in the midst of an open space of streaming winds and mists. This cubic foot of air certainly exists. As a mathematician computing the amount of air in this space I may deal with such cubic feet and nothing else. My attitude and concepts may imply that these cubic feet are the sole and total fact; but of course they are not. Furthermore we cannot say that nature has set apart these cubic feet as of peculiar significance; she has not enclosed with natural partitions nor in any wise given them prominence. It is scientific method that gives them unique significance and makes them shine forth with unnatural light. Only in this sense is the world that concerns science artificial. It is unnatural because it gives a prominence which nature does not give to certain features, and deals with certain portions of total fact as though they were the whole when they are but a very small part. It is in this sense that the world of practice and theory, of common sense and science, is a "theoretical construction."

This scientifically defined world is inimical to beauty, love and worship only on two conditions: First, if we fail to see that it is but a part, and a very small part at that, of the total fact of experience; and, second, if we assume that it is more important and worthy of consideration (except for scientific purposes) than the rest of that which we experience. We must see that over and above these conceptually defined molecules and atoms and vibrations, and round them and through them, there flows that total event of nature which enters awareness in the form of the concrete fullness of experience. We must see that this streaming flood of fact is not of necessity any less significant or worthy than those features selected for scientific treatment.

But the worst evil arises when we separate these two parts of total fact and, in our effort to do justice to both, treat them as independent worlds. All too frequently we have considered that experience which is aesthetically appreciated, loved and worshipped as constituting a wholly different world from that which is scientifically defined, controlled and predicted. It is this opposition between what some have been pleased to call the realm of values and the realm of facts that we wish to designate by the two metaphysical demands. Our conviction is that there are no two such realms or worlds outside our own fancies. The two are one. If the humpty dumpty of total fact were indeed broken in two we could never get him together again. But the great fall and break has never occurred except in the form of a nightmare which we have dreamed, and are now unable to put out of our minds.

We demand love and beauty and worship; and we also demand prediction, intellectual clarity and efficient control. Those who construct their metaphysics with only the religious interest in view, or only the scientific, in either case do great damage and are equally far from the truth. For these contrasting demands of life are equally valid; both are derived directly from experience; both refer to indubitable data; both are rooted in human nature; and both rear their constructions in faith and aspiration. An adequate view of the universe can be attained only by learning from both and giving equal consideration to both.

Since science and religion are both derived from immediate experience, their metaphysical assumptions cannot be inconsistent if rightly formulated. But of course it is exceedingly difficult to rightly formulate them. To the degree that they are inconsistent we know that they are not altogether correct; and there generally is some inconsistency between them, which only means that we have never yet developed a perfectly correct metaphysics and probably will not yet for a long time. Mere consistency between them would not prove them correct; but inconsistency proves them wrong. The development of a correct metaphysics must be the work of many generations yet to come as it has been the work of many that are gone. It is not an impossible task; it can be advanced from age to age. But it is one of those tasks which may be still incomplete when human history is done. To have a share in it is no mean thing.

Just now the work of metaphysics requires a reconstruction of the basic concepts of science on the one hand and of religion on the other. This work of reconstruction must be done chiefly by specialists in the two respective fields. After they have done their work the philosopher may examine these concepts in relation to one another to see wherein they are inconsistent or consistent and point out what further work awaits to be done. He may himself contribute something to their mutual modification, with the view to attaining consistency.

The basic concepts underlying the appreciative side of life and requiring modification, are those of purpose, personality and beauty. They are at the present time undergoing development as are also the underlying concepts of science. We shall try to trace some of this reconstruction and rapprochement in the chief ideas that underly these two interests.

METAPHYSICAL RECONSTRUCTION AND RAPPROCHEMENT

This movement toward a more adequate total view of the world, due to change in the underlying thought of these two aspects of life, results in part from the development of the more concrete sciences, such as Biology, Psychology, and Sociology. The influence of Biology in particular should be noted. Biology[3] is still not sufficiently founded to resist as fully as it might the transcription of its terms to those of physics, one of the oldest of the sciences. But Biology is rapidly becoming able to establish its own distinctive categories. Lloyd Morgan,[4] L. T. Hobhouse,[5] J. B. S. Haldane[6] are among

---

[3] Significant is L. J. Henderson's Work, *The Order of Nature.*
[4] *Emergent Evolution.*
[5] *Development and Purpose.*
[6] *The New Physiology and Life.*

those who have drawn upon Biology in the interests of metaphysics.

Another movement working to the same end of recasting the concepts of mind and purpose versus matter and mechanism, so that a more adequate metaphysics may arise, is the theory of emergence.[7] According to this view matter does not exclude life, nor life mind, nor mind some higher entity, such as God.   Life is not something left over when one has analyzed out all that constitutes matter, nor is mind the remainder that is left when one subtracts what constitutes matter, and life.   Rather, matter, when it has attained to a cerain degree of complexity in organization, gives rise to a wholly new quality called life, just as chemical components, when brought into a certain form and complexity of organization, give rise to something wholly new—water out of hydrogen and oxygen, for instance.   The water does not cease to be hydrogen and oxygen, but at the same time it is something radically different from them.   The new relation into which these elements enter is not merely an additive relation but a creative one.   Water can be analyzed into hydrogen and oxygen but can never be properly called "nothing but" hydrogen and oxygen.   So also a living organism may be analyzed into inorganic elements in such a way as to leave nothing over.   But these elements organized in a certain way constitute something wholly different from inorganic matter;

---

[7] Typical exponents of this are S. Alexander, *Space Time and Deity;* and Lloyd Morgan, *Creative Evolution.*

namely life. The new relation into which the elements enter is a creative one.

Of course it does not necessarily follow that the more complex levels of existence are chronologically later than the more simple, although it is thought that this is the way they have evolved on our planet. Whether or not there ever was a time when there was nothing but matter throughout the whole universe, only later giving rise to the emergence of life and other higher levels of existence, is purely a matter of speculation. The best we can do is to trace the evolutionary process on our own planet, and even that is very incompletely known. And of course the story of our little earth is by no means the whole story of matter, life and mind. not to speak of God. So far as concerns the logic of the theory of emergence, there may have been always the more complex organizations along with the more simple; the universe may have contained throughout all time those complex systems which are required for the existence of the higher orders.

We do not wish to set up this metaphysical view as the last word of truth. We are only pointing to it as typical of a movement now going on which renders obsolete the old, exclusive materialistic assumptions of scientific thought, and shows the influence of Biology and the new theory of relativity in recasting the prevailing views of the universe.

But perhaps the most radical and important recasting of these basic concepts of science is to be traced to the theory of relativity itself. While this theory is not a metaphysical doctrine, it quite completely breaks down certain assumptions of an earlier Physics. The best exponent of the wider applications of relativity is A. N. Whitehead. While he has not himself taken up the positive work of metaphysical reconstruction, he has done the negative work of showing that the metaphysics of materialism is impossible. How closely he treads on the borders of metaphysics and how wistfully he gazes into those mysterious regions is indicated in the following statement:[8]

> . . . . the past and future meet and mingle in the ill-defined present. The passage of nature which is only another name for the creative force of existence has no narrow ledge of definite instantaneous present within which to operate. Its operative present which is now urging nature forward must be sought for throughout the whole, in the remotest past as well as in the narrowest breadth of any present duration. Perhaps also in the unrealized future. Perhaps also in the future which might be as well as the actual future which will be. It is impossible to meditate on time and the mystery of the creative passage of nature without an overwhelming emotion at the limitations of human intelligence.

When he speaks of the creative force of existence; when he says that the operative present includes the remote past and may also involve the future, and the future that might be determines the

---

[8] *Concept of Nature*, p. 73. In a work published since this was written, *Science and the Modern World*, Whitehead develops the metaphysical implications which we have here detected.

operative present as well as the future that shall be, is he not coming very close to saying that mind and purpose are at work in all nature? When we come to formulate a concept of purpose and mind, as we shall do forthwith, I think we shall find that this description fits it most remarkably. If to understand the operative present we must search the future that might be as well as the future that shall be, we certainly have mind and purpose. Of course Whitehead only throws this out as a suggestion, but a suggestion which is altogether compatible with the basic concepts of physics and the other sciences.

We must now turn from the metaphysical assumptions of science to those of religious experience. For if it is necessary to recast the basic concepts of science in order to develop an adequate total view of the world, it is not less necessary to recast the basic concepts of religious experience. We refer particularly to the concepts of mind, personality and purpose. The Christian religion is based on the assumption that personality and purpose must somehow be at the root of all things just as much as the exact sciences are based on the assumption that nature involves predictable and controllable elements. But the concepts of personality and purpose which have prevailed in the past have been most unsatisfactory—just as unsatisfactory as the concepts heretofore prevailing in science concerning that in nature which is predictable and controllable. Just as the theory of relativity has forced a change of these scientific concepts, which is favorable to a better adjustment of science and religion, so also it is necessary to bring about a change in our concepts of purpose and mind if we are to meet this new advance half way and enable these two sides of life to work together in that fashion which is so important to the good of human life.

We believe the great difficulty experienced heretofore in trying to find mind and purpose in the world, even in human beings, in such form as clearly to think and describe it, has been due to the very inadequate concepts of these objects of search. We cannot find an object if we do not know what its chief characters are. We must know what it is we are looking for. The kind of purpose which some have tried to find in life and in the world generally is a kind of "purpose" which has no existence anywhere. Naturally it could not be found.

Let us turn to a rigorous scientific thinker and arch-positivist for light on the nature of purpose, mind and personality. He is by no means the only one who has presented this view of purpose. It has been independently developed elsewhere. But since he has allied himself with science and considers the beliefs of religion pure illusion, the view of purpose which he considers valid can scarcely be inimical to the claims of science. We refer to E. Rignano.[9]

The living organism is distinguished from the inorganic, according to Rignano, by what he calls "mnemonic accumulations." In the highest intel-

---

9 He has developed his ideas in *The Psychology of Reasoning.*

ligences, such as that of man, this may appear in the form of memory, recognition, and that supplementing of sensations which gives rise to perception, as well as in other intellectual processes. But in the lowliest organisms, and indeed wherever life is found, we find these mnemonic accumulations in the sense that the living always strives in such a way as to regain or preserve the conditions to which it has become habituated. If it has become accustomed to a certain temperature, it is restless until it regains that temperature. In other words, past happenings control the behavior of the organism; while the inanimate thing, like a stone, is controlled by present happenings only. The stone does not try to regain the temperature to which it has become habituated, nor the degree of light which has become customary to it. The organism has the peculiar characteristic of "reactivating" some past experience, of "reproducing it" or having it again "evoked," so that it becomes a dominating factor in the operations of the present.

Thus life means the accumulation of past events. It means the amplification of the present moment by integrating it with the past. For matter as such the present moment is the only time there is. The material particle is not controlled by reactivation or reproduction of what has befallen it in the past. For it the past is nothing; the present, only, is. But anything which is capable of forming habits is controlled by something more than the present; the past still lives with it. This is true of "blind" habit; but it is also true of conscious purpose, of planning and thinking, of aspiring and loving.   In fact this property of mnemonic accumulation, if taken broadly enough, seems to be at least one necessary factor to account for all the phenomena of life, such as growth of the cell from embryo to adult, the reproduction of maimed portions of the organism, the healing of a wound, the whole process of anabolism.

When it comes to the higher "spiritual" forms of life it is plain that love and aspiration require that past experience be drawn upon and that it continue to operate in controlling the behavior and providing material for thought and contemplation. As we saw in the previous chapter love, more than anything else save worship, gathers up the past, contemplates it in the present and by means of it foresees the future.   May we not conclude, then, that life, purpose and spirit all have this generic character of mnemonic accumulation.   At the levels of what we ordinarily call spirit and purpose it certainly attains a breadth, complexity and form which is not found at the lower biological levels. But it is a further amplification and development of this generic character.

Here, then, do we not have a tenable concept of spirit and purpose?   It is the preservation of the past as operative in the present and as shaping the future.   Such mind and purpose throughout nature would be an "operative present" which included "the remotest past as well as the narrowest breadth of any present duration," and furthermore included "the future that might be" as well as the "future that shall be." So far as Biology has investigated, this peculiar "mnemonic accumulation" has always involved physical organism. Naturally Biology could not find it elsewhere for that is all that Biology investigates. But the biological organism may be only one instance of this. In fact, Whitehead seems to be very sure that the whole passage of nature involves in its operative present all the past.

What the total fact of this world may be we can not clearly and fully state. Recent reconstruction of the basic concepts of science have served not to reveal the nature of this total fact but to mark out the limits of scientific knowledge and to remove certain old metaphysical assumptions inherited from the Greeks, which have hindered the assimilation of the scientific to the religious view of the world. What we do know is: (1) that certain finely devised experiments yield certain results with a very high degree of uniformity, and (2) that a certain awareness of the concrete fullness of experience yields values indispensable to human living. With these two legs to our ladder we can climb. Without both we are undone. The metaphysician is free to climb to the top of this ladder and find what is there to be seen. But let him not kick the ladder from beneath his feet. Let him not think he can dispense with either leg. Experimental verification on the one hand, and the values of immediate experience on the other, are the only supports he has and the only supports the human race can have. These are the two sides of life which we have been portraying.   We must never allow our inordinate interest in the one to destroy concern for the other.   For the one without the other comes to nought.

CHAPTER VI

SCIENTIFIC METHOD

Science does not deal with all that is. It adopts the method of Descartes and attacks the universe piecemeal. The scientist isolates certain portions of the world and deals with them to the exclusion of all else. "He assumes, for instance, that in considering a small portion of the universe he can neglect all the rest. He goes on this assumption until he finds it is wrong. If it is wrong, he looks round and brings another little bit of the universe into his ken, and continues altering his field of observation until his isolated system behaves as though it were really isolated. All the time he is able to leave the whole universe as such alone; he gets all the advantages he could have got out of a theory of the universe without the disadvantages."[1]

Any such portion of the universe selected by science for treatment is not, necessarily, a unit of time and space. It may be a fine thread of correlation which can be traced throughout time and space.

A most excellent statement of scientific method is made by Ralph Barton Perry: "Scientific description, then, is governed by two motives, on the

1 Ritchie, A. D., *Scientific Method,* p. 7.

one hand, unity, parsimony, or simplicity, the reduction of variety and change to as few terms as possible; and, on the other hand, exact formulation. When a scientific description, satisfying these conditions is experimentally verified, it is said to be a law. . . ."

Professor Perry proceeds to set forth the way in which Galileo formulated the law describing the acceleration of bodies falling to the earth. Galileo ignores everything about the body and its fall save only certain very specific properties of time and space. He describes it all in terms of d, which stands for the constant variation in distance from the starting point, and t, which is the corresponding constant variation in time. The fact that the body passes me as it falls, or that it reflects the light of the sun, or that it crushes a flower when it hits the ground, and innumerable other features about the fall which may be just as striking to a casual observer and have more immediate vital significance to individuals at the time, all this is excluded from consideration as being irrelevant. Irrelevant to what? Irrelevant to the particular kind of concept Galileo was trying to formulate. And what was the nature of this concept?

Here we come to the central motive of scientific investigation, the index of its great value and the mark of its severe limitations. Galileo was trying to formulate a concept of falling bodies of such a nature that it would exclude from consideration all the unforeseeable, innumerable, indefinable, uncontrollable, swarming multiplicity of features that enter into any such fall, but would clearly specify certain abstract features which could always be foreseen, calculated and accurately described, in any such fall. And why was it so exceedingly important to abstract just these features from all the rest? From the standpoint of pure science (and this is the dominating motive of the scientific investigator despite the effort to blur the distinction between pure and applied science) the value of such a concept is that it enables the investigator to perform *experiments in imagination,* to combine in thought innumerable operations, thus saving the time and labor involved in physical experimentation, and above all making it possible to perform experiments on nature far beyond the reach of physical experimentation. One can perform experiments in thought and combine them in many different ways, and calculate the results with certainty, only if he can isolate in such experiments some one feature which will certainly be present in all cases and will certainly fulfill itself according to a mathematical formula. That is just exactly the sort of feature in falling bodies which Galileo succeeded in isolating. Hence the great importance of the concept which he formulated.

Let us quote further from Perry's account of this conception of the falling body with its distance covering property *d,* and the time of its duration *t.*

> Now in this elementary mechanical conception of uniform acceleration, appear all the most essential principles of exact science. It is a de-

scription of motion, because it simply records the behavior of the falling body, and does not seek further to account for or justify it. It is an analytical description, because it expresses motion as a relation of the terms, such as *d, t*, etc., into which it can be analyzed. It is an exact description because the terms and relations are mathematically formulated. And it is a simplification and unification of phenomena, because it has discovered a constancy or identity underlying bare differences. As we proceed to more complex concepts we shall not, I think, meet with any new principles of method as fundamental as these. . . .

I am led to conclude, therefore, that all of these concepts are essentially ratios or relational complexes of the simple terms of experience, such as space, time, color, sound, etc., and that each of these ratios or relational complexes expresses some specific complexity or configuration, which is found in nature. And I judge that these concepts illustrate the motive of science; which is simply to describe and record, with special reference to their unity and constancy, the actual changes of bodies.[2]

We have quoted Professor Perry at some length because we think this is a most excellent statement of the way science portrays the world. Also it serves to introduce a matter of prime importance.

Professor Perry says that science describes in the simplest terms possible with accuracy, exactly what occurs in a certain mass of experience made up of space, time, sound, color, etc. But the matter of

2 Perry, R. B., *Present Philosophical Tendencies*, pp. 55–62.

prime consideration to us just now is the fact that that same mass of experience could be described with equal accuracy in wholly different terms and the object which that description would disclose in the event experienced might be wholly different from what the ordinary simplified scientific description portrays. Take, for instance, our solar system. The accepted scientific description of it is the Copernican-Newtonian, because that is the simplest. The object it presents is the central sun with the planets moving regularly and ceaselessly about it in elliptical orbits. But that same identical experience which we all have when we turn our eyes to the sky at night could be described with equal truth by a modernized form of the Ptolemaic description, in which the planets do not move about the sun but in complex orbits of their own. It would be correct, furthermore, to describe the sun as moving about the earth. The difficulty with such descriptions other than the Copernican is not their incorrectness but the vast complexity that they involve. They are unscientific not because they are untrue but because they would be so complex as to be almost useless for further scientific purposes.

Here, then, we have a given experience of the sky. Following Whitehead we can call this fact experienced an event. This event may be correctly described in many different ways. How many different ways we do not know. Each different correct description would reveal in the event a different object. We have already noted two different correct descriptions revealing two very different objects in the same event. How many other correct descriptions there may be of this event, and consequently how many different objects may be "ingredient" in it, we do not know. There may be any number of different objects in it, differing in character from one another far more than the two we have mentioned. And these other objects are just as truly existent as are those which science does treat. Most of these objects science deliberately ignores for good reason. They are too complex.

So far as the claims of science are at fault, perhaps the greatest cause for maladjustment between science and religion has been the failure of the followers of science to see that science treats of only a very few of the objects that inhere in experience.

The fallacy of simplification consists in assuming that, when a natural mechanism has been traced out and described the phenomena characteristic of it have been *fully* set forth and accounted for. As a matter of fact, however true and valuable the mechanical description may be, it is after all an account of part of the facts, not of the whole of them. To assume the contrary is to forget that a mechanical description is necessarily in general terms, and that the more immediate and peculiar features of the object under inquiry are abstracted from and not included in it. A description in terms of adjustment, motion, number, order, and so forth, is the kind of description that will apply to multitudes of cases, and which therefore must omit what is distinctive of those cases taken singly. Especially does it fail to do justice to the data, its interest being in the relations, not in the elementary facts.[3]

It is a little difficult to illustrate the many different objects that inhere in a given bit of experience, because science searches for that one object which is most simple and can be made to serve her purposes best. And since science is the only means we have for searching out with accuracy and verification the different objects that enter into experience, we have difficulty pointing out these other objects. But a little consideration makes it plain that these other objects are truly existent there. For instance, matter is described by physics in terms of electrons and matter is certainly an electronic object, if this view has been completely verified. But the object we deal with in terms of electrons is very different from the ordinary perceptual object of matter. The various sense qualities do not enter into the physicist's description. Yet there certainly is a material object made up of these sense qualities or to which these qualities pertain. A description that leaves them out, simply leaves them out. It is not dealing with the same object, although it may very well be dealing with the same event or mass of experience.

A thought similar to this is expressed by A. B. Eddington, who writes:

> We have a world of point-events with their primary interval-relations. Out of these an unlimited number of more complicated relations and

---

3 Cooley, W. F., *The Principles of Science*, pp. 149, 150.

qualities can be built up mathematically, describing various features of the state of the world. These exist in nature in the same sense as an unlimited number of walks exist on an open moor. But the existence is, as it were, latent unless someone gives a significance to the walk by following it; and in the same way the existence of any one of these qualities of the world only acquires significance above its fellows, if a mind singles it out for recognition. Mind filters out matter from the meaningless jumble of qualities, as the prism filters out the colours of the rainbow from the chaotic pulsations of white light. Mind exalts the permanent and ignores the transitory; and it appears from the mathematical study of relations that the only way in which mind can achieve her object is by picking out one particular quality as the permanent substance of the perceptual world, partitioning a perceptual time and space for it to be permanent in, and, as a necessary consequence of this Hobson's choice, the laws of gravitation and mechanics and geometry have to be obeyed. Is it too much to say that the mind's search for permanence has created the world of physics?[4]

Perhaps the most striking contrast of different objects entering into the same space-time complex appears when we consider how the different sciences deal with a human being. He may be described as a dance of electrons, and the description would be correct. He is that sort of object to the physicist when the latter is employing a certain technique. The whole man, without residue, could be

---

[4] Eddington, A. S., *Space, Time and Gravitation*, p. 196.

analyzed into chemical terms. A different object is then brought to light. Again he may be reduced to the terms of biology and we have still another object in the same event. Again the psychologist and sociologist describe him and he becomes still another object. Finally the man's personal friend discerns in him another object, his friend, who is quite different from a swarm of electrons, or a system of bio-chemistry, or even a sociological entity. This object with which the friend deals cannot be so accurately described and verified as can the scientific objects which inhere in the mass of experience, because science is the only means of accurate description and verification, and this object which concerns the friend is so intricately complex as to escape the reach of any scientific method now in sight. But no one will doubt that this friendly object exists. He is an "object of belief" rather than an "object of scientific theory," but he does not become any the less real on that account. But still on beyond the object to which the friend reacts, is the object which the man discerns when he considers himself. His own introspective and self-conscious view of himself reveals still another object.

A given event, to use Whitehead's language, may have an infinite number of different objects as ingredients in it, according as it is described in one set of terms or another, or as it *might be* correctly described, whether any one actually does so or not. But all the sciences put together have not yet begun to bring to light all the objects that inhere in most events. Furthermore, they never will. They never will for one thing, because they seem to be limited to the most simplified descriptions. Also many correct descriptions would be of no value to science. Science will develop only those descriptions which can serve the purposes of science. And the purpose of science, we have seen, is to formulate definite concepts of those features of experience which can be certainly predicted, accurately calculated, and experimentally treated in imagination, thus transcending the narrow limits of constant physical experimentation, and opening the way to the discovery and correct description of innumerable objects otherwise completely hidden from us. Of course physical experimentation can never be discarded, but it becomes merely a check and a guide to the intellectual combination of great numbers of experiments, most of which do not have to be performed by actual physical manipulation at the time because similar experiments have been performed at other times, and on the basis of these it is possible to calculate what the result will be if the conditions are given. To experiment in the imagination means to set up certain imaginary conditions and calculate what the result would be if certain operations were carried out. One of the chief means of scientific discovery is by just such imaginary experiments. For instance, a certain result is observed. The investigator sets up certain imaginary conditions and calculates from them what the result will be. If the result thus attained by calculation from imaginary conditions proves to be identical with the result under physical observation, he has strong reason for believing that the conditions he has imagined are the actual conditions producing the phenomenon under investigation. Of course this is only one step in the total process of discovery, but it is one of the most important.

The point is, that science consists far more in the purely imaginative and intellectual performance and combination of experiments than it does of physical manipulation, however indispensable the latter may be. And this imaginative and intellectual treatment of its subject matter is only possible because of these predictable and accurately described features of experience which have been brought to light by such concepts as those formed by Galileo concerning falling bodies. It is no longer necessary to experiment physically with every falling body to see whether in its case it is true that $v/t=g$. Because it is the purpose of science to discover features of experience which can be subjected to such purely imaginative and intellectual treatment, the following statement by E. Mach, concerning scientific description of fact, holds true: "A rule reached by the observation of facts, cannot possibly embrace the entire fact in its infinite richness, in all its inexhaustible manifoldness; on the contrary, it can furnish only a rough outline of the fact, one-sidedly emphasizing only the feature that is of importance for the given technical or scientific aim in question."[5]

---

[5] Mach, E., *"Die Mechanik in ihrer Entwicklung historisch-kritisch dargestellt."*

Here we find very narrow limits imposed upon science in its efforts to portray the world in which we live and the objects with which we deal. We do not point to these limitations with any sense of gratification, for we are all losers thereby. We certainly should like to discover the scientifically verified truth concerning all the objects with which we deal. But we cannot do so because the very accuracy and precision which science introduces into knowledge forces her to use only the most simple terms—terms so simple that they cannot describe these more complex objects.

Any mass of experience is actually all those different objects which it might be correctly described to be, whether any one ever has the wit and skill to describe it so or not. Furthermore, in undergoing that mass of experience, we are actually experiencing all those different objects. We may not have any knowledge of them. We certainly have no knowledge of most of the objects that enter into the manifold of ordinary experience. For instance, we do not ordinarily cognize the electrons that enter into daily experience, nor the light of the stars that shine by day or the invisible stars at any time, nor the currents of air that flow about us, nor the minerals that compose the soil beneath our feet, although all these do enter into our experience in the sense of shaping the total occurrence that we experience. Scientific discovery consists in bringing to light some of these ingredients of our experience. Among these objects which enter constantly into our daily lives are doubtless many which no science has ever yet guessed at. Into such daily experience there constantly enter other human minds, our associates. Into such daily experience there also enters God, however unnoticed, because it is this total event which yields the religious experience when our awareness is sufficiently opened to it. The total occurrence which we experience is actually all the objects which it may be correctly described as being. Such being the case it is plain that we are constantly and intimately in converse with innumerable objects which are wholly beyond the ken of the several sciences.

Some of these objects are known to us by "common sense" and are objects of belief, as Perry would say, but are not subject to scientific description. Such beliefs canot be tested by science, either because of the complexity of their object or because such objects of belief do not fit into the purpose of scientific description as that purpose was above defined. For instance, the total personality whom I love and contemplate, including that particular tone of voice and glint of eye and pressure of hand and much else of rich, concrete, sensuous material, cannot be scientifically described for both reasons just mentioned.

Psychology and sociology come nearer to describing the beloved personal object than any other sciences now at our command. But even they fall far short of it. Furthermore, the degree to which they seem to approximate a scientific description of that concrete fullness of experience which enters into the beloved personality, is probably illusory, for they can seem to approach this concrete fullness of description only by abandoning the scientific method and falling into what is more or less the method of pre-scientific appreciation. In fact, psychology and sociology have hardly yet attained the status of science, and are constantly falling into the inaccuracies and guesses of pre-scientific thinking. Let us quote from J. B. S. Haldane on this:

> Why then am I not a psychologist? Because, with all respect for psychologists, I do not think psychology is yet a science. Mechanics became a science when physicists had decided what they meant by such words as weight, velocity, and force, but not till then. The psychologists are still trying to arrive at a satisfactory terminology for the simplest phenomena they have to deal with. Until they are clearer as to the exact meaning of the words they use, they can hardly begin to record events on scientific lines. Moreover I do not believe that psychology will go very far without a satisfactory physiology of the nervous system, any more than physiology could advance until physics and chemistry had developed to a certain point. This is not to say that physiology is a mere branch of physics or chemistry, or mind a mere by-product of the brain. But it is a fact that we can only know about life by observing the movements of matter. You may be the most spiritually minded man on earth, but I can only learn that fact by seeing, hearing or feeling your bodily movements. As the latter

depend on events in your brain, I may as well get some information about those points. To study psychology before we understand physiology of the brain is like trying to study physics without a knowledge of mathematics. Physics is more than mathematics, as matter is more than space, but you cannot have one without the other. Now at the moment the physiology of the nervous system is being worked out with great speed, and by contributing to its progress I suspect I am do-doing more for psychology than if I became a psychologist.[6]

But if the apparent closeness that psychology seems to attain in describing the object that I know in friendly intercourse as my friend, be due to its lack of strict scientific method, it is probable that when psychology does become strictly scientific it will so delimit its field and simplify its terms that the object it discloses will be much more remote from that common sense object that I know as a personality. The behaviorists, for instance, would seem to be among the most scientific of the psychologists, but the stricter of them deny that they have anything to do with consciousness, or mind, or personality, as these are generally understood by common sense. It may be claimed that there are no such objects as consciousness or mind or personality in the sense in which common sense understands them. If in taking such a position one means to assert that the definitions and descriptions by common sense are inaccurate, then one is certainly

---

6 *The New Republic*, Dec. 3, 1924.

in the right. For common sense is not accurate. If one claims that there is no such thing as color as understood by common sense, meaning thereby merely to assert that the definitions of color made by common sense are inaccurate, one is in the right. But if one means to assert that there is nothing at all but electromagnetic vibrations, because these descriptions are scientifically accurate, and that there is no color at all that could be correctly described in any other terms, then one is certainly in the wrong. The same is true of mind, consciousness and personality.

The question is this: Are these objects which common sense inaccurately describes, but which it is trying to describe, and which could not be described save by terms much more complex than any psychology or other science could possibly employ, truly existent in nature independently of human ideas about them? Our answer must be that many beliefs of common sense are blundering attempts to think correctly about objects which we experience and which are truly existent in the sense that an omniscient science could portray them, but which are beyond the reach of our present science.

It is in this extra-scientific way that we know God. We are acquainted with certain masses of experience which have as an ingredient that which is more important for the safety and growth of human life than anything else, although we are unable to specify just what it is. That object is God, because the word God signifies such an object, whether known or unknown.

We have already described in different ways that particular state of awareness which has God as its object. We have spoken of the mystic awareness of the total datum in which all data are merged. But since we have referred to the matter just now in relation to the limits of science, let us describe this state of attentive awareness in terms of one of the keenest and most thorough scientific thinkers of our time. A. N. Whitehead states precisely what we mean by such a state of awarenss. He does not give it any religious significance, but he sets it forth as altogether valid. He writes:[7] "The immediate fact for awareness is the whole occurrence of nature." But this whole occurrence of nature rarely engages our attention. We do not ordinarily cognize it, have any emotion with regard to it, or react to it in any way. Rather, what engages our attention, what stirs our emotion and awakens us to response, is some entity or factor which we have discriminated within this total event. It is these objects ingredient in the event which are the matters of prime concern' to practical and theoreical interest as they operate in ordinary life or in scientific investigation. But, nevertheless it is true, as Whitehead says, that:

"The ultimate fact for sense-awareness is an event. This whole event is discriminated by us into partial events." [8]

But it is possible for attentive awareness to be opened to this total event. For the most part, it

7 The Concept of Nature, p. 14.
8 loc. cit. p. 15.

is true, we consider only "partial events." "We are aware of an event which is our bodily life, of an event which is the course of nature within this room, and of a vaguely perceived aggregate of other partial events. This is the discrimination in sense awareness of fact into parts." [9]

But there are times when this whole occurrence of nature, this total event, may engage attention. "Occasionally our own sense-perception in moments when thought-activity has been lulled to quiescence is not far off the attainment of this ideal limit." [10]

Whitehead does not seem to attach any value to this state of awareness. He suggests that "The sense perception of some lower forms of life may be conjectured to approximate to this character habitually." That is, of course, a possible conjecture. But we believe all the evidence is to the contrary. It would seem that the awareness of the lower animals is very rigidly determined by certain automatisms and that these constantly hold the attention upon certain individual data or, as Whitehead would express it, certain factors, which are necessary to guide the behavior of the organism. However, it is, of course, pure conjecture to talk about what the lower animals may be aware of. If they are aware of "the total fact" which is inclusive of all nature, still it would not be for them a religious experience. For in order that it be religious, one must derive from it those values which religious experience

9 *loc. cit.* p. 15.
10 *loc. cit.* p. 14.

yields, and the lower animals show no evidence of this. This state of awareness certainly does not come to us humans in ordinary sleep or semiconsciousness or stupor. At such times the attention is engaged by fragments of sensuous experience that are revived from the past and stream more or less chaotically before the mind in the form of phantasy. To become aware of total fact is a rather difficult art, although some mystics seem to have a natural aptitude for it and there are fleeting moments when perhaps we all approximate it more or less, as Whitehead suggests.

This "total event" can no doubt be correctly described as having electrons as its ingredients, and many other things besides. It can also be described as having God as its ingredient. That is to say, it can be correctly defined as that which, when made the object of attentive awareness, yields the values of religious experience. That is what we mean by God when approached from the standpoint of experience. This immediate experience of the "whole occurrence of nature" may signify time and space and color and sound, for all these are ingredients of nature; but it also signifies God. And its divine significance does not exclude the others, nor the others the divine.

Mr. Whitehead is very careful not to discuss the religious and metaphysical implications of his concept of nature. Especially does he insist upon rejecting what he calls the "bifurcation" of nature, which means that the object or event observed must not be confused with the mind that does the observing. Because the mind describes nature, the nature thus described must not therefore be thought to partake of the nature of mind.. In this we most heartily agree. We feel that he has contributed immensely to the clarification of the whole field of experience by this insistence; and that he has opened the way to a far more satisfactory interpretation of the object of religious experience as well as to a more satisfactory scientific description of natural processes. As long as we think that the observing mind must perforce give its own qualities to everything which it observes, our thinking is constantly moving in a circle and we find ourselves headed back toward solipsism with almost every turn. This view, that whatever we know must partake of the nature of mind, has seemed an easy way to prove the postulates of religious faith. If everything we know must partake of the nature of mind, if everything for which we strive must partake of the nature of an ideal which is, by definition, a spiritual entity, then it seems easy to prove that we have constantly to do with a supreme mind like our own, a universal idealizing personality like unto ourselves. But the way is altogether too easy. It proves too much. If we follow that path rigorously to the end, and not merely just to that point where it seems to serve our religious needs, we find ourselves at last merely with our own mental constructions. Everything we know partakes of the nature of self. Self closes us in on every hand. We are caught in the net of self, and struggle as we may we can never deal with anything save the self, for everything must forthwith turn into the self as soon as we have any dealings with it at all. Such a magic circle is intolerable. We may take the leap of faith, of course, and claim that there must be that outside the self which acts upon us, but it is a blind leap of faith. This way of thinking has lured too many into its toils. Whitehead breaks free of it and opens up the wide regions of fact independent of self and clears the suffocating air laden with too much self.

But this view of nature presented by Whitehead and others, which insists that nature does not partake of the nature of self merely because the self cognizes it, and experiences it, has been greatly misinterpreted.[11] This view most certainly does not imply that nature must therefore be mindless or that a universal mind may not inhere in nature. Whitehead simply does not enter into that problem, but there is nothing in his theory of bifurcation of nature which contradicts that view. For the bifurcation of nature applies to the confusion of self with objects observed; it implies nothing beyond that concerning the nature of the observed object. Nature may very well be moved and sustained by the operation of a supreme mind or personality. In fact, Whitehead is most emphatic in his rejection of materialism; and we believe that his work, together with that of others akin to him, has rendered materialism altogether obsolete among the well informed. Furthermore, his insistence that the event, rather than the object, is the basic fact of nature and that nature is marked by a "creative

---

11 *See*, for instance, Edgar Pierce, *Philosophy of Character*, chap. IX.

advance" lends itself most readily to the religious interpretation. For "event" and "creative advance" mean precisely that in nature the past does not drop out of existence with each successive instant, as materialism would declare, but that the past continues operative with the present to shape the future. And this, as we saw in last chapter, is precisely the best concept we can form of mind, purpose and life. Events pass, but the point-moment is not the only kind of event. It is not even the most important event, except for purposes of scientific analysis. It is no more "real" or ultimate than many other kinds of events. There is, above all, one all-inclusive event including all other events, and having as its ingredients all objects. Not all events involve a mind, but some do; and it may be that what gives the character and creative advance to the whole of nature and every part of nature is that there is operative throughout the whole of nature a Mind.

While Mr. Whitehead, as already noted, is very careful not to discuss the religious and metaphysical implications of his views, we believe his thought would not be inhospitable to what we have just written. Our reason for thinking so is based on a quotation which Mr. Whitehead makes from Dean Inge; and also Whitehead's brief treatment of life and mind in his *Principles of Natural Knowledge*.[12]

---

12 Chap. XVIII, on "Rhythms."

The quotation from Dean Inge follows:

> Dr. Inge's paper is entitled "Platonism and Human Immortality," and in it there occurs the following statement: "To sum up. The Platonic doctrine of immortality rests on the independence of the spiritual world. The spiritual world is not a world of unrealized ideals, over against a real world of unspiritual fact. It is, on the contrary, the real world of which we have a true though very incomplete knowledge, over against a world of common experience, which, as a complete whole, is not real, since it is compacted out of miscellaneous data, not all on the same level, by the help of the imagination. There is no world corresponding to the world of common experience. Nature makes abstractions for us, deciding what range of vibrations we are to see and hear, what things we are to notice and remember.
>
> I have cited these statements, [continues Mr. Whitehead, referring to another previous quotation along with that from Dean Inge] because both of them deal with topics which, though they lie outside the range of our discussion, are always being confused with it. The reason is that they lie proximate to our field of thought, and are topics which are of burning interest to the metaphysically minded.

In his treatment of life and mind under the head of rhythms in *Principles of Natural Knowledge,* he says that an individual life is not an object. "To say that the object is alive suppresses the necessary reference to the event; and to say that an event is alive suppresses the necessary reference to the object."[13]

Life means rhythm. That is to say, it constantly recovers a certain status. But this rhythm is not altogether uniform. It passes out of the given status in various ways and recovers it in various ways. In other words, a life learns with the passage of time, or, it synthesizes the past with the present to some degree.

Putting the matter in other language, life is that which learns. It is that which establishes habits and acquires new habits.

We only refer to these statements from Whitehead to indicate that this view of nature, which must constitute the assumptions of science henceforth, according to Whitehead, gives ample space for all that religious experience demands.

To summarize: Scientific method is a device by which we disencumber ourselves of the great mass of experience in order to pick our way through the intricate space-time structure of all experience. The quickest, easiest, and surest way to pass from one object to another in thinking, especially when one is dealing with a great number and variety of objects, is to ignore everything about the objects save their space-time structure and to consider only certain distinctive features of that. Now, of course, all the sense qualities of experience cannot be ignored. We cannot pick our way through some phase of the intricate structure of experience unless we have at least a few sense data to guide us. But

---

[13] *loc. cit.* p. 196.

science endeavors to reduce these data to the utmost simplicity and to those particular data which serve to guide in making inferences, but which may not give any of the sensuous flavor and fullness of the experienced object. Science sifts out the great mass of experience, winnows it, and breaks it up until it discovers those rare and scattered bits of data which serve to guide inference through the maze of the space-time structure. These bits of experience called data, which are thus selected by microscope and spectroscope, by mortar and acid and many other devices, these data are made to shine like jeweled lights at strategic points and critical turns of the space-time structure. But the world thus plotted and planned by science presents a wholly different aspect from that which is known by way of ordinary experience. The total concrete object, which can be known with some degree of accuracy by means of common sense, is not the object which engages the attention of the scientist. He has not time to deal with such a cumbersome mass of experience.

Let us illustrate. Science has a concept which designates not the experience of heat, but that motion called energy, or the space covered by the mercury in a thermometer. There is no particular resemblance between the rising and falling mercury and the experience of heat and cold. Science does not designate the experience of a falling body, but the ratio of motion which is the distinguishing feature of the space-time structure. It designates not light as experienced but the vibrations, which again is space-time structure. It designates not the experience of sound, but the vibrations of the air, etc., etc. So it is that science develops a technique for dispensing with the massive bulk of experience, dealing only with certain carefully selected data, the ultimate precipitate of analysis. All else is cast aside as irrelevant.

What we have said of science in general is just as true of psychology and sociology as it is of physics and chemistry. It is not so manifest in them partly because they are more complex, partly because they have not yet developed an adequate technique, and partly because they are so young that they are not yet thoroughly scientific in their thinking. We do not mean that all science must be reduced to the terms of physics and chemistry. On the contrary, we are very sure that each science must develop its own viewpoint, its own method and technique; and each must interpret the world in its own categories. But since all experience has the form of space and time, all science that studies experience (and science can study nothing else), must work out the complex space-time interrelations of those data of experience which it has selected as its own peculiar subject matter.

Scientific method transforms the character of our experience because it transforms our habits of response. The stimuli that once aroused us no longer stir us in so far as we assume the scientific attitude, for the scientific attitude means to be responsive to certain rarified and selected data at certain loci in space-time. The massive bulk of experience is ignored. In so far as we are scientific the concrete object with all its savor and rich sensuous fullness and emotional stir is gone. It has disappeared like a bubble. Compare the world as known and felt by the modern scientific man with the world of the primitive man as revealed in his folk lore. To be sure this latter world is full of fancy, illusion and error. But all this efflorescence of illusion and emotion springs from a rich concrete sensuous mass of experience which he has not learned to ignore. The child has it; the poet has it; the great originative scientist has it who breaks with the scientific tradition of his day and, like Darwin or Galileo, blazes new paths through the jungle of experience. But the established scientific technique does not have it and can not have it, because of its very nature.

Now the question we want to raise is this. Can any such vast and elaborate scientific technique, as we see developing in the civilized world of today as never before in history, exist in our midst without greatly transforming our appreciation, our habits of response, our capacity for growth, for joy, and for fullness of life? Certainly a technique rapidly accumulating from generation to generation, and becoming ever more pervasive in shaping our habits, appreciations, and outlooks, can not be without effect. What effect will it have? The answer to that question has already been given. The effect it must have is revealed in the very nature of scientific methodology or technique. It impoverishes the world of experience. It destroys the capacity to appreciate the "surplusage" of experience.

What does that mean?   This surplusage of experience means all those data, and all that unanalyzed datum, which is irrelevant to our established scientific theories, but is indispensable to the development of any radically new theories and the solution of very different and more complex problems that are bound to arise with an evolving civilization. It means, furthermore, all that concrete and scientifically meaningless, experience which makes seven-eighths of the joy of living—a lover's kiss, the colors of the sunset, a child's soft little hands, the heart of a rose.

Yes, but, you answer, our scientific technique has not destroyed in us the capacity to enjoy the bits of concrete experience just mentioned. Of course not. We could not illustrate our point with experiences we had ceased to appreciate. The only point is this: Does scientific technique have any effect upon human life? And, if so, what effect? It seems there can be but one answer. It vastly magnifies our efficiency in procuring anything we want and have the capacity to enjoy; but it greatly diminishes our capacity to enjoy. It gives us wonderful instruments of achievement, but narrows and distorts our vision of what is to be achieved.

But there is art and religion to correct this one-sided development of science. Exactly. The more science advances, the more elaborate, constraining, and pervasive its technique becomes, the more we require art, religion, and the personal intercourse of love to save our souls. But while science has advanced with leaps and bounds since the Middle

Ages, since the days of Greece and Rome, since the days of Thor and Brunhilde, have art and religion kept pace with it? Are art and religion more or less pervasive throughout our life in comparison with those days? Is scientific technique more or less pervasive?

Are we asserting our case is hopeless? Not at all. We are simply pointing to what we believe is a manifest fact which must not be ignored and can be corrected if we face it. We must turn to the cultivation of art and, above all, religion for salvation.

CHAPTER VII

AWARENESS AND SCIENTIFIC DISCOVERY

The first step in widening the range of knowledge is to open the mind to fact. By fact we mean not objects but the passage of nature, the undiscriminated occurrence with all its merged qualities and undistinguished relations. The first step is not to open the mind to objects because that cannot be done. To discern objects is the last step, not the first, in widening knowledge. We cannot discover new objects until after a process of theorizing and experimentation. We must form theory after theory until we have hit upon the right one, and we must perform experiment after experiment until we get the one that will adequately test that right theory. Only then do we discover the object with scientific certainty. We cannot, therefore, open the mind to objects at once. But we can to some degree open the mind to sense awareness, which is the first step in acquisition of knowledge by scientific method or by the methods of common sense.

But it is by no means a simple and easy matter to open the mind to novel sense awareness. It is probably as difficult, and certainly as important, as anything else in the total process of scientific discovery. It is sometimes thought that all we need to do is to stare in order to become aware of the whole movement of nature in that particular quarter to which our perception is turned. But such is not the case. Our observation of an event is cribbed and cabined to the narrowest limits by the traditional concepts which shut us into a recognition of only those few scattered objects in time and space which the demands of biological welfare and the cultural history of our people have forced upon our attention. The rest of that total happening which is going on about us all the time, we ignore. Our minds are shut in by a vicious circle. We ordinarily cannot become aware of anything save what these traditional concepts define for us; and we cannot ordinarily develop new concepts without becoming aware of something new.

This difficulty of widening the bounds of awareness must not be confused with another difficulty which has received much more attention—the difficulty of thinking. It is difficult to theorize with caution, to learn to devise adequate experimental tests and to withhold judgment until the tests have been sufficient. That is the difficulty of scientific method; that is the difficulty of transforming awareness into knowledge. But there is the prior difficulty of becoming aware of anything new. We must first observe that which will suggest a problem before we can employ our powers of thinking for its solution. It is possible to have great power and technique in thinking, but no awareness of novel fact upon which to exercise this power and technique. The scholastics of the Middle Ages give us an example of this. The Greeks themselves were prone to this weakness, although they were probably more receptive to concrete fact than any other people of their time. The scientific technique of any time may be threatened by this limited range of awareness.

There are rare individuals whose awareness breaks free of the bonds of tradition to a degree much beyond that of others. The scientific genius and discoverer is not only a great thinker. He is more. He is aware of happenings which others ignore. The mystic may not be a thinker at all, but if he is mystic in the sense we have defined the term, his awareness of fact may be immense. The artist and the aesthete has a widened and sensitive awareness. The prophet, the seer, the lover, are aware of more than the ordinary hard-headed practical man who is solely concerned with fitting into the mechanism of the social system.

To make new, revolutionary discoveries, to turn human investigation of nature into new regions, as Thales did, as Copernicus and Kepler and Newton and Galileo and Pasteur and Darwin and Harvey did, one must take note of that which lies round about, but which we have not been taught to observe by the established theories of science nor by traditional common sense. We do not mean to say that such open awareness will itself alone lead to the discovery of new objects. It is only part of the process of discovery. But it is one indispensable part. However helpless awareness may be without concepts, concepts are equally helpless without awareness. This teaching of Kant still holds true. But when open awareness is combined with the disciplined habits of scientific investigation, we have those flashes of insight that lead to the theory of gravitation, or the heliocentric solar system, or that bodies fall with the same speed regardless of weight.

But when open awareness is not combined with trained and efficient powers of thought it not only fails to yield new knowledge, but may destroy efficiency in making adaptation to environment. For efficiency in any given social system generally requires that one react in the ordinary way to those objects which society in general recognizes. If any one's excess of awareness causes him to ignore these objects, or to react to them in a different way, he is likely to become socially inefficient with respect to the impersonal social adaptation.

The aesthete and mystic are sometimes of this sort. Even when such persons have retained their practical common sense, their widened awareness has not always enabled them to contribute anything of value to the common good, except the good of testifying to the possibility of this wider awareness. This living testimony of the mystic, however, may itself be a great good and he has sometimes been esteemed for this alone. But he has also for the same reason been an object of contempt. If the aesthete is an artist as well as an aesthete, if he is not only widely and richly aware, but is also master of the trick by which to transmit his awareness to others, he may, with good fortune, receive his social reward. But no doubt many aesthetes are not artists. Furthermore, this creative work of the artist is very different from the creativity of scientific discovery. The artist does not necessarily add anything to human knowledge nor help others to do so. If no theory is applied to those added regions of experience which he brings to light, and if these forms and qualities are not relevant to any theory that is likely to occur to mortal mind because of the bias given to human minds by the demands of biological welfare and the past history of culture, then this increment of awareness is worthless to all scientific discovery, however illuminating it might be to minds with a different biological bias and cultural history. But that certainly does not mean that it has no human value. To be more widely aware, as the aesthete and the mystic are, is a great human good. It is the independent good of art and religion. Let not the practical and theoretical mind deride it, lest they destroy the good of their own practice and theory.

Since the mystic and the aesthete have often had no concepts with which to define their experience, cognize it, or communicate it, their raptures have sounded like raving foolishness. Being unable to contribute any good to the common welfare, save the good of their own enthusiasm and rapture over that which enters their awareness, they have sometimes been looked upon with contempt. But do they not testify to the value of that which is immediately given to those who have eyes to see and ears to hear?

This living testimony of the religious seer, with his rhapsodies and incoherent utterances, has sometimes rendered him very precious to certain simple folk who are glad to know that there can be such awareness even though they themselves cannot enter into it. But such simple folk are attracted to the mystic generally because they themselves have the glimmerings of such a widened awareness, enough to make them realize that other eyes may see what their dim eyes cannot behold. They are glad to know that there is something beautiful or holy here in their midst, even though it be withholden from their own awareness. And they are simple folk, often, precisely because they have just enough of this widened awareness to distract their attention and interest from those practical pursuits and scientific theories which lead to social success.

Now, of course, any awareness that cannot be communicated and tested will offer a great opportunity for tricksters and quacks, and these have no doubt often availed themselves of the opportunity. But the very nature of simple awareness makes this inevitable, and the existence of such quacks cannot disprove the genuineness of such awareness on the part of some. Indeed, it is because more or less of this awareness exceeding the bounds of common sense is so wide spread that these quacks have so much success. The people who are duped are often those who have just enough of the vision in themselves to make them readily believe one who claims to have more of the same thing. But are these "dupes" to be pitied more than the hard-headed individuals who never glimpse anything in heaven or earth, save those few clear cut bits of natural furniture which common sense has so definitely set up as the whole world in which we live and amid which they have learned to live with almost as much comfort and stolid contentment as the beasts. It is not they who are likely to be deceived by a false prophet, but those who have glimpsed that which the prophet claims to see.

This awareness of which we speak must not be confused with idle dreaming or phantasy. The most natural thing for the ordinary human is to dream, which is neither thinking nor awareness, but a muddle of the two. In this state we do not open the mind to the fullness of immediate experience, but rather our awareness is obscured and distorted by the reverie. Awareness is sometimes called intuition. Rignano speaks of the part it plays in scientific discovery. "....in case of really new discoveries concentration is as a rule an impediment. In this domain intuition and chance are always the reigning powers and the unexpected flash of insight of a genius is of greater value than untiring reflection and pertinacity."[1]

Further on he says:

> But if we find that intuition as opposed to "attentive" observation or "pondered reasoning," has the drawback of being much more subject to danger of error, it has nevertheless a much greater probability of yielding entirely new truths. Clearly what is necessary for these new

[1] Rignano, E., *The Psychology of Reasoning*, p. 127.

truths to appear is first "imagination," that is, a mind that is capable of bursting through the boundaries of ordinary associations. An observer, for instance, not possessed with this power, never would have perceived in a swinging lamp anything but qualities that are quite ordinarily perceived, the material of which it was made, its shape, the carvings which adorned it and such like things. But even Galileo himself would not have discovered the isochronism of the oscillations if he had not just at this moment luckily been seized by an affective preoccupation concerning the measurement of time. This preoccupation, was, as far as it concerned the lamp, of a totally new kind, and it would not have been born if Galileo, stimulated by one of the ordinary affective interests relating to the lamp (an artist's æsthetic interest, a verger's interest in making sure that it had not gone out, or that it was not too dusty) had been examining it with attention. Similarly the playful imagination of a Faraday needs to give itself an entirely free hand, without being checked at every turn by the incessant limitation, exclusion and control, exercised by both the affectivities of a strong state of attention.

Thus we see not only the usefulness, but the absolute necessity of a continual alternation between intuition and reflection. If this latter has need of the former to escape from the grave danger of sterility; intuition in its turn has need of reflection to control and prove the validity of each of the new observations or discoveries that it makes or thinks it makes at each new free flight of imagination.[2]

---

2 *loc. cit.* pp. 129, 130.

Human creativity consists in bringing together these two sides of discovery, open awareness on the one hand and theorizing on the other—with its analysis, discrimination, definition and experimentation. When these two are united and rightly balanced, human life leaps forward like an open spillway or a hound unleashed. Life becomes suddenly and marvelously abundant. When these two are brought into fruitful interaction, the richness of the world and the fertility of life is shown to be amazing. The artist, the prophet, the moral and social reformer, the scientific genius, the religious seer, all rise up in numbers and power when awareness of the wide, rich, novel fullness of concrete experience can be combined with the scientific method. But wide open mysic awareness flounders helplessly and blindly when unassisted by scientific method. And scientific method becomes a barren definition of concepts without yielding anything to enrich life when not supported by open awareness.

So we come back to our original question: How can open awareness be cultivated? Often it is not cultivated at all, but is a gift of nature, and in extreme cases a mark of genius. But it can be cultivated and is cultivated in the forms of art, love and worship. It cannot, as we have seen, be attained by concentrated attention and methods of efficiency. But it can be, and is, attained in the ways mentioned. The awareness attained in art, love and worship may have no value for scientific theory. That depends upon how the two are adjusted to one another. But even when open awareness has no value for practice and theory, it has great human value. Few would question the value that proceeds from awareness of the beautiful, the beloved and the divine. Science needs to be supplemented and counterbalanced by these three, even if they did nothing directly to contribute to scientific discovery. But they may also contribute to scientific discovery under favorable conditions.

We shall make a brief study of aesthetic and religious experience in order to show how these cultivate awareness, and to further show the nature and value of this awareness and its rightful place over against scientific method.

In aesthetic experience we find a more or less vc̓uminous flow of sensuous awareness. In music there is a flow of sound, the rhythm serving to keep this stream of immediate experience within the bounds of awareness. The stream may be analyzed into distinct data, but in aesthetic appreciation these data form a unity or continuum. If there is a complete cessation of the sound it is aesthetic only when it serves to accentuate the continuity of the experience that has just been going on and will shortly be resumed, thus making more vivid and comprehensive one's awareness of the unity. The breaks and variations are only sufficient to hold the experience continuously and in its wholeness within awareness.

Playing over the surface of the sound, or merged with it, there may be fancies, ideas, even intricate logical processes. The music may suggest these. We believe generally there is some such free play of fancy. But the body of the aesthetic experience must be this stream of immediate awareness; and nothing else than this is required for a genuine aesthetic experience. There may well be ideas but they must suggest and maintain the flow of sensuous experience, not suppress it from awareness as they do in ordinary discourse.

But art consists just as much in excluding from awareness a large portion of the total occurrence of nature as it consists in bringing to vivid awareness that particular portion which constitutes the work of art, for instance the music. Art is thus both negative and positive in opening the mind to the events. Artistic genius consists just as much in excluding certain elements from consciousness as in admitting others. It would seem that this work of exclusion is so important because otherwise awareness could not be held constantly receptive. If anything else were admitted to the magic unity it would dissolve and break up into the data of ordinary selective attention which admits only those meager features which are of practical significance. The stream of sensuous experience would fade out, leaving only those discrete elements that signify the objects of common day. Art, then, must be very highly selective, but in a manner exactly opposite to that of ordinary life. Ordinary selective attention has been built up by traditions of common sense within such narrow limits that it is now difficult to recover any awareness of continuous experience, except by means of skillfully wrought works of art. There is that much truth (and error) in the oft heard statement that beauty is the vision of reality; it is our awareness of some event in its wholeness before it has been broken up into separate elements and considered piecemeal as the analytic mind is prone to do.

Religious experience differs from the aesthetic in that it is not awareness of some work of art. Of course, works of art may be used to bring on the worshipful attitude, but awareness must include more than the artistic production; it must be receptive to some more inclusive event, ultimately to that totality which is the "operative present . . . urging nature forward" including "the whole, in the remotest past as well as in the narrowest breadth of any present duration." Of course aesthetic experience may be directed upon some aspect of nature and not a work of art. But it is more limited in scope than is the religious awareness. There may be, however, all degrees of transition from the aesthetic to the religious and back again. There are aesthetic experiences that border on the religious, and religious experiences that border on the aesthetic.

We have spoken of music as our example because, in a certain sense, it can be called the purest form of art. But the same principles apply to other forms. In painting the selected portion of experience brought to awareness is a blending of colors in such form that they can be held in awareness as a single continuous experience. This awareness is generally much enriched and made more vivid if the proper ideas are suggested, and in most paint-ing such ideas are suggested. But if the ideas be-cloud, break into, or distract from appreciation of the sensuous fullness of that which is presented in the art, the appreciation is not aesthetic and we miss' the art in so far as it may be there.

In poetry we have a still more complex form of art. Here the stream of experience must be brought to awareness, first by reviving masses of past ex-perience, and, secondly, by delivering these masses to awareness as an indivisible unity. This is done by the suggestiveness of words and rhythm. Here ideas play a larger part than in music or painting. When masses of experience must be revived from the past, ideas are necessary to a degree that they are not when the experience can be provided in the immediate environment. There seems to be no other effective way of reviving past experience, ex-cept by means of ideas. But the ideas must always be subordinated to the function of sustaining this stream of revived experience before attentive con-sciousness. On the other hand, this stream of ex-perience may well give a pungency and otherwise unwordable significance to the ideas. But always it is art only as there is this massive body of sensuous experience presented to awareness.

De La Mare's *Listeners* will serve as an example of such art.

> "Is there anybody there?" said the traveller,
> Knocking on the moonlit door,
> While his horse in the silence champed the grasses
> On the forest's ferny floor,

And a bird flew up from the turret
Above the traveller's head.
And he smote the door a second time:
"Is there anybody there?" he said.
But no one descended to the traveller.
No head from the leaf fringed sill,
Leaned over and looked into his grey eyes,
Where he stood, perplexed and still.
But only a host of phantom listeners,
That dwelt in the lone house then,
Stood listening in the quiet of the moonlight,
To that voice from the world of men;
Stood thronging the faint moon-beams in the
    dark stair
That goes down to the empty hall,
Hearkening to an air stirred and shaken
By the lone traveller's call.
And he felt in his heart their strangeness,
Their stillness answering his cry,
While his horse moved cropping the turf,
'Neath the starred and leafy sky.
For suddenly he smote the door again, even
    louder,
And lifted his head.
"Tell them I came and no one answered,
That I kept my word," he said.
But never a stir made the listeners,
Though every word he spake
Fell echoing through the shadowiness
Of the lone house, from that one man left awake.
Aye, they heard his foot on the stirrup,
And the sound of iron on stone,
And how the silence surged softly backward,
When the plunging hoofs were gone.

Here we have concrete experience brought to our awareness by reference to certain objects, and this experience is presented as a unity, partly by means of the rhythm, but more by means of a vague idea that runs throughout, too obscure to divert attenion from the stream of experience, but sufficiently suggested to preserve the continuity of the whole. Above all, the idea is so presented that one seeks it in the suggested stream of experience rather than in the words or propositions of the poem. This is what the art of poetry requires.

This balance between idea and stream of experience is not often so artistically perfect as in this poem. It is very common for the idea to overbalance in the sense that one seeks for it in the verbal propositions; and the idea may be so clearly and logically developed, rather than merged in the suggestive symbolism, that the art fades out almost altogether. In other cases there may not be sufficient idea to revive and hold the stream of experience before awareness, but only a crackle of words.

Something akin to both aesthetic and religious experience, perhaps midway between the two, is love. When two lovers are in one another's presence, especially in the early stages of unsophisticated love, there sometimes comes over them an overwhelming sense of simply being together. Their immediate awareness of this event is so absorbing, and it is so compact of concrete fullness, that they cannot think, analyze or discriminate. Simply to experience the total fact involved in being together is all that they can ask; and to appreciate that in its fullness requires all the powers of their minds. There is a flood of experience involved in this being together, which may include much revived from the past as well as the immediately present. There is scarcely any distinct object cognized or purpose defined or motor attitude dominant. If the behaviorist insists that there must be motor attitudes, we would reply that probably there are so many of them confused and merged that it is as though there were none at all for all practical purposes. There is no thought, or scarcely any, and little or no purpose; there is only that vivid awareness of the rich volume of fact involved in being together.

The experience of the mystic in his communion with God is similar to this.

So we have in the immediate experience of beauty and of love that which is closely akin to religious experience. The latter differs from the other two in being awareness of a more inclusive event. In the experience both of beauty and of love, awareness is receptive to a wealth of experience far greater than that of ordinary life; but in both definite limits are set. In the aesthetic it would seem the limits are narrower than in the erotic. But in both the exclusion of incompatible elements is just as important as the inclusion of compatible; and in art especially is it plain that the technique is just as much concerned in excluding as in including. But in religious experience no such fixed limits are set, or none which we can be sure of. It is true that a certain religious tradition may confine religious awareness to certain fields of experience, or seem to do so, but it is much more difficult to impose these limits upon the individual. In art these limits are very plainly prescribed because the experience is of certain definite and technically constructed objects. In love the experience, while not so clearly defined, centers about one single definite human being with certain specific characteristics; and while the experience of love is not merely of the other human organism but may include innumerable features of the whole environment, both present and past, nevertheless the whole experience centers in, and is controlled by, one's awareness of this other human individual with all his limited and definite characteristics. But in religious experience there is no such limiting and defining object. In religious experience there are of course limits to the range of one's awareness, but these limits are set by our own mental nature and not by the nature of the objects as is the case in beauty and love. Religious experience is much less clearly defined and may be much richer in undiscriminated content. On the other hand, however, since its form and limits are not clearly defined, it may be much poorer in content. In other words, it may fluctuate without limit and without the constraint of any form. However, for the ordinary person, religious experience can scarcely be aroused, except by the presentation of some more or less definite form. Hence the indispensability of religious tradition, with its symbols, ceremonies, doctrines, institutions, distinctive philosophy, ethics and art.

This awareness, which we say constitutes religious experience, should not be confused with emotion. Of course, there is likely to be emotion in it, just as when we experience a tree—"woodman, spare that tree"—or when we experience the "old oaken bucket." But that does not mean that the emotion is the same identical thing as the color, shape, coolness, etc., which we call the bucket. We experience certain sense data which enter into the constitution of the bucket or signify the bucket. The emotion is not a datum in the same sense as the color, shape, etc. There is plainly a distinction between the data of experience and our emotional reaction to those data.

Now this distinction holds just as truly when the experience entering awareness does not consist of discreet data, as in ordinary perception, but consists of that continuous flow of sensuous experience we have described as aesthetic or mystical. In the mystical experience all data are so merged that one can scarcely speak of any perception at all, but only of an undefined awareness. In such an experience it may be impossible to distinguish introspectively between the sensuous experience and the emotion, since neither is defined. But that there is a distinction can scarcely be doubted. For, of course, the distinctions in experience hold whether we recognize them or not.

Now when we are aware of this sensuous flow we are likewise, in many cases, aware of an emotion, although that is not necessary. But merged with the emotion, and possibly indistinguishable from it, is this flow of qualitied space-time signifying

God to the mystic. The emotional reaction to this datum must not be identified with the datum itself. In other words the event which is being experienced must be distinguished from the process of undergoing the experience, which latter might be called the percipient event, after Whitehead.

There are several reasons why the datum of religious experience should be so commonly considered subjective, or confused with emotion. In the first place, it is not analyzed and broken up into separate data and set in a framework of space-time relations, as our ordinary objects of perception are. On the contrary it is an experience of the continuous flow of space-time with its innumerable qualities merged into an undifferentiated mass, not because the discreet qualities are lacking, but because these are not analyzed out, but allowed to continue in their natural original state of mergence with many other qualities, most of which we have never learned to discriminate.

In the second place, our awareness of this flow of experience is ordinarily very dim, fluctuating, fading off into total unconsciousness, and therefore the emotional reaction is proportionately much more vivid and distinct as compared to the other elements in the total experience. Consequently, in retrospect it will be the emotion rather than the other data which will come to mind, especially since there are no distinct and discreet data to be held in memory.

In the third place, and most important of all, there is no well defined concept of that which is experienced, and hence no means of clearly thinking or cognizing the object ingredient in the total event experienced.

As emotion has been given an undue place in religious experience, so has the subconscious. Let us try to define the place of the subconscious in it.

When we become aware of that total fact which yields the religious experience, we respond. This response may include an uprush of previously suppressed impulses. This may give rise to an efflorescence of phantasy, vision, photisms, contortions, and illusions, along with occasional flashes of profound insight and visions which may later be demonstrated to contain much truth. For instance, the truth that we can overcome evil with good and can conquer our enemies with love, may well have been discovered in such flashes of insight. This exuberance of the imagination is found in the records of many mystics.

While the source of this fire works of fancy may be certain impulses hitherto suppressed, the important question for us is: What releases these impulses from the latent or suppressed state? What we experience religiously is not these suppressed impulses, but it is that which so stimulates us as to arouse these latent impulses. The significant thing in religious experience is that object which produces such a response. This uprush from the subconscious is an incidental by-product. What we experience is not primarily the subconscious, but it is that which produces this response of the subconscious.

We feel there is no more dangerous misinterpretation of religious experience than to represent it as "subjective." Our whole point has been to show that it is an experience of something not ourselves. Our reason for connecting it with the experience of beauty and love has been to show that the same problem is involved in them; and it has seemed to us that a consideration of the three together would help to show that in each there is something not ourselves which we experience.

In appreciating music the beautiful sound is not something "inside the soul." It is just as much out there in the world of space-time as is the locomotive whistle that warns us to get off the track. We may react to the locomotive whistle with all sorts of inner spasms, tensions, and glandular processes, yielding a tumult of emotion. But the sound with all its qualities is there in the world of sense. The same is true of that which we experience in love and in worship. In these latter we may not so definitely cognize the object signified by the datum, but the object is "out there" just as much as the locomotive and its whistle.

It has been said that art is a flight from reality; that it does not reveal the world of fact, but leads us into a fictitious realm where we can satisfy ourselves with dreams. There are two senses in which this may be taken.

In the first place, in so far as the beautiful object is a construction of human wit and so a remaking of things as they are, it is a flight from things in their raw state to things as artificially constructed. But in the same sense a bridge or railroad or a stone wall is a flight from reality, since it is a transformation of nature from its raw state. But I think that is somewhat of a *reductio ad absurdum*. The beautiful thing, when constructed, is now a fact made of clay or wood or sound or color or what not; and its beauty is there to behold as much as the bridge is there to use.

But a great deal of beauty, by far the greater portion of it, we believe, is not a construction of human wit. It is nature in its rawest state; a sunset or a mountain peak, for instance. Aesthetic appreciation in such cases is the most immediate confronting of reality and accepting of it. It is the most complete exposure of our sensitivity to the impact of that which is.

But in this respect the religious experience exceeds the aesthetic, for the object of religious consciousness is never properly the construction of human wit, as the beautiful object often is. In religious experience we confront and accept the immediate deliverance of all our sensitivity; we expose ourselves most completely to the impact of fact in its rawest and most massive form.

There is, however, a second sense in which it can more truly be said that art is a flight from reality; or rather there is a form of art which is so. There is an art which is sentimental and romantic. Its beauty consists not so much in that which is presented to the senses as in the fancies it suggests. There is a great deal of that sort of "art," especially in recent times. But it is a great mistake to think that all art is of this kind; and certainly a still greater mistake to think that all beauty is of this sort.

What has been said of beauty applies also to love. There is a sentimental love which is blind, which is a flight from reality, and little else than phantasy. It consists in dreaming of an "ideal," loving that ideal and using a human person to symbolize that ideal without any great regard to the true nature of the individual so used. But such sentimental love is not the only kind there can be. There is a love which consists in response to the total individuality of the beloved. The beloved opens the gates of awareness to the lover, and the latter becomes aware of a concrete fullness of fact that could never enter his awareness if he did not love.

When we come to religion we find also these two kinds, as we do in the aesthetic and in love. There is a "religion" which is flight from the facts of experience and there is a religion which is the most complete opening of awareness to the total fact of immediate experience.

But there are many defenders of religion, by no means sentimentalists, who insist that the experience is "within," and not a matter of sense at all. Many mystics are very emphatic in such declarations. We believe this seeming divergence from the view we are presenting arises from a difference in the usage of words. Sense has been used to designate the data of selective attention by which we perceive the objects of ordinary life. Hence when

attention does not thus select it is thought that
sensation does not apply. Certainly sense data
apprehended as signifying the ordinary objects of
daily life is not the usual way of experiencing God.
But when we follow up the teaching of the mystics
we find that their experience is a matter of sense,
if we understand by sense that sort of awareness
we have been describing.

As an example of the assertions of a mystic who
denied that sense is a way to God, let us consider
the statements of Rufus Jones. In speaking of religious experience he says: "But there is no denying
that we are moving now in another field from that
of sense-facts and we cannot get the same coeffi-
cients and common denominators that mark group
experience in the well-known frameworks of space
and time."[3]

It is certainly true that mystic experience is very
different from "group experience" and the perceptions of the herd with their "well-known frameworks of space and time." This well-known
framework of space and time consists of certain
abstract features carefully selected from out of total
fact by the experimental methods of common
sense working through centuries of transmitted tradition, and now at last set up as the total world of
sense objects, when in truth they constitute only
dots and dashes discriminated from the total happening. This well-known framework of space and
time consists of chunks of empty spaces, regions
of nothing and regions of something. And its

---

[3] *Fundamental Ends of Life*, p. 114.

sense-facts are selected from the stream of experience because of their utility in guiding the operation of certain established and useful habits. But to say all this does not mean that the mystic experience has nothing to do with sense and space-time. On the contrary three short pages further on from the above quotation we find Rufus Jones saying what is most typical of the statements of all mystics and implies all that we have said about their experience.

"We are these strange eternity-haunted beings just because we are conjunct with God whom some of us at least discover walking with us in the cool of the day, as the fish feels the ocean or the bird the air."

What more intelligible account of such an experience can be given than the fusing of the total sensitivity of the organism in awareness of that which is given in immediate experience! But Rufus Jones, like so many mystics, claims that we cognize self, other human selves, and God in some other way than by sensuous experience.

> We not only look out upon the objects which occupy space, and deal with them through ten special senses (we used to say five) but we further discover by an "inward ho" that we have an inside existence of our own—an inner self. We come upon it by a flash of insight which our senses can not explain. . . . Not only so, but we also by a similar act of the unified inner self acknowledge and appreciate spiritual centers in the persons around us. In doing this we vastly transcend all reports

of our senses. We overleap everything we see and touch . . . The mystic does not stop with these two ways of transcending sense experience . . . He holds that it is possible to have direct, first-hand experience of God—the Spiritual Presence in whom we live our own spiritual lives and who is the potential environment of all personal selves. It is possible for the sensitive soul to feel the pulse-beat of the Eternal Heart . . . The mystic claims that the human soul is bosomed on the deeps of a spiritual sea of Life which flows around it, and that the sensitively adjusted life can catch intimations of celestial currents and can gain clues and hints of direction, even when the cruder senses make no reports and give no guidance.

Now all this we feel to be profoundly true and it can be accepted on its face value with our interpretation of the mystic experience. But to deny that sense and space-time have anything to do with his experience of God is to throw all these claims of the mystic into a region where modern thought and scientific method can make no connection with them. No matter how conciliatory toward science the mystic may be, he is irrevocably estranged from all scientific method as long as he makes this denial. Mysticism and scientific method need one another for reasons already stated. But as long as the mystic insists that his experience has nothing to do with sense and space and time, the two can only gaze at one another across a great abyss.

CHAPTER VIII

REBIRTH VERSUS AUTO-SUGGESTION

Philip Cabot, in his recent book, *Except Ye Be Born Again*, recounts his own experience of rebirth. He tells of his failing health, worries, failure to discern any value in the continuance of life, yet grimly clinging to it. He tells of retiring to a solitary place and spending many days, reading and thinking until there gradually came to him a sense of the divine presence, the annihilation of old habits, interests, purposes, views, a great sense of rest and peace, and the emergence, out of this state, into a wholly new way of life.

He expressly contrasts this experience with that form of prayer which is auto-suggestion. He does not deny great value to this latter. Indeed, he says that this is the most common, and is indispensable to right living. He notes the fact that from time immemorial religious people have prayed early in the morning and late at night; and it is known that at these times subconscious processes are most readily accessible to conscious control. He says:

> The miraculous cures that have been accomplished by disciples of mind-cure, Christian Science, and auto-suggestion, seem to me to result from a method which is in fact common to all

of them, although it has been obscured by superficial differences which have been over-emphasized.
Each of them has developed a formula or method
by which the mind of the patient is concentrated
on the conception of health, at times and in ways
which successfully transfer this image to the seat
of action in the subconscious.   This concentration
is the secret of their success, and I am tempted to
believe that the miracles of healing of all times
rest upon the same foundation.   The simplest and
perhaps most effective example is the formula
of M. Coué, repeated twenty times night and
morning.

Now it is impossible for me to doubt that if
the same concentration can be achieved in Christian prayer, similar but more far-reaching cura-
tive results will be produced.   I hold that the
great problem for each of us in developing the
technique of prayer is to ascertain exactly by what
method such concentration upon the symbols of
his faith can be produced in his individual.

But he goes on immediately to contrast all this
with that experience of God which is "the deepest
form of worship." This deeper form of religious
experience involves what he calls "The Seeing
Eye" and the "Listening Ear." Now this seeing
eye and listening ear make up that distinctive form
of religious experience which we have been trying to
describe. In it we cease to respond to special stimuli
but become responsive to the undifferentiated mass
of innumerable stimuli that are playing upon us.
We cease to distinguish and interpret special selected
data but react to the concrete fullness of immedi-

ate experience, which is constantly pouring over us and through us, but which we ordinarily ignore.

We believe it is exceedingly important to clarify the distinction between this "deepest form of worship" and any form which, like auto-suggestion, seeks the fulfillment of already established wants. What this deeper form of worship does is not so much to satisfy our pre-existent wants, but to transform our wants. When this transformation is sufficiently radical we call it rebirth. And this we claim is the chief function of religion in human life, immeasurably more important than its service in helping to fulfill the wants which we bring to it. This function of rebirth has been too often ignored in studies of religion. Religion is treated as a device by which we satisfy our wants. But when it transforms our wants, it does just the opposite; it makes it impossible for us to satisfy our pre-existent wants, even to that degree that we were able to fulfill them by our own unaided efforts. It renders those old wants worthless. Of course, if we seek the experience of God with the wish to be transformed, then that want is fulfilled. But that is very different from what is ordinarily understood by fulfillment of wants in prayer.

Now this transformation of wants, this rebirth, is precisely what auto-suggestion cannot do, no matter how Christian or unchristian it may be. In auto-suggestion we have certain wants which we release from inhibition and develop into efficient and persistent habits by the practices which go by that name. But these practices are successful just in proportion to the definiteness with which we know what we want. Even when we say "getting better day by day" we are seeking to satisfy certain established wants, the fulfillment of which constitute what we mean by getting better. And this most general formula is discarded for more specific statements whenever an individual case reveals the nature of these special wants.

There is a kind of prayer, then, which can be brought under the general head of auto-suggestion, the state of mind which arises when we enter into the experience of God rendering auto-suggestion much more effective. It is plain that the breaking down of habitual attitudes and the release of free impulses in that diffusive stimulation of the total personality, which occurs in religious experience as we have described it is precisely the condition in which new habits can be formed and the impulsive resources of the individual organized about certain dominant wants. This is akin to auto-suggestion; and it is no less religious when called auto-suggestion, if it is practiced in that unique form of experience which is distinctively religious.

But it must be noted that this breaking down of established habits and attitudes and this melting down of the whole personality into a free play of impulses in response to the total mass of stimuli playing upon the individual, if it goes far enough, will wipe out those very wants which the individual brings to the hour of worship for fulfillment. For a want is a motor set, a more or less effectively organized system of impulses. In that deliquescence of personality which occurs in the deeper reaches of worship, these may be melted down with all the others. As the individual rises up out of such a profound experience he may arise with a new self, meaning a system of impulses organized in a manner radically different from that which constituted his personality before the experience. And this reorganizaion of impulses means a new outlook on life. It means attention to features which he previously ignored.

But the chief value of such profound religious experience is that it makes possible a unification of the personality. Systems of impulse, complexes and habits, which have hitherto been at cross purposes or dissociated from one another, are now fused into a single total system of impulse. It means that our habits, instead of being antagonistic or indifferent to one another, are now fully coöperative. It means that the individual no longer must work against inner, mental friction. It means that he is made whole, according to the good old Christian term. And this unification of impulses fused into a single total system in which each element sustains every other, is blessedness, peace and power.

This melting down, fusing and unifying of all impulses and habits, is something which the growing personality requires periodically. It cannot be done once and for all, because new impulses and habits are constantly arising; and as these multiply in adaptation to changing features of the environment, they are bound to come into more or less conflict with one another. Suppressions and dissociations will occur. The only way in which mental health can be preserved is by periodic melting down of these motor sets, and out of this deliquescence of habit into free impulses a new and unified system of response arises.

Of course, when we speak of the melting down of habits into a free play of impulse, we speak in relative terms. How far the established character of the individual is resolved is a matter of degree. There are always left those nuclei of habit around which the new system must be organized. If we could fix on those wants which are identified with these basic nuclei of our total individuality and about which all our lives must be organized if they are to be organized at all, then we might say that even this deepest kind of worship is a means for fulfilling our wants. Then worship becomes a process by which we dig down to the most basic wants of all human life and restore these wants to their rightful place as basis and center for the organization of all other wants.

But this we cannot do by a purely meditative, intellecual process. We cannot do it intellectually, for one thing, because we have not yet been able adequately to define the deepest wants of human living. There are many proposed definitions, of course, but there is no scientific verification that these statements of the basic human wants are correct, and they contradict one another. But more important still, it cannot be done by any such purely intellectual process, because even if we did know what we most deeply wanted, we still could not resolve our established system of habits to such a degree as to make possible their reorganization about these most basic wants. It is religious experience alone which can do this adequaely. Other experiences do it to a degree, for instance the aesthetic and erotic. But none can do it sufficiently for the needs of life, save only the experience of God.

But to clarify this religious function of rebirth, we mus give one or two instances of it. In the *World Tomorrow,* issue of December, 1923, is a careful and accurate record of a woman's experience, written by herself, from which we quote. It is anonymous, but we believe altogether authentic.

> My own longing to fall in love was partly gratified when I was twenty. I partly loved a young man of about my own age. He only partly loved me; and although we made a little love to each other, it was subject to chilling doubts, and recoils that overcame both our hearts, though not always simultaneously. I remember one summer afternoon lying alone on our cliffs and knowing a pure and perfect rapture of the body. But such a perfect sweetness seldom came; and I did all I could to kill with sentimentality my honest, sweet anæmic little emotion. . . .
>
> The war was the terrier that shook the whole rat of my life. From the time the soldiers went marching across the border to Mexico I was a pacifist. Even amid the fever of jingoism that seized upon the people here in the Southwest, I had been a pacifist, accordingly, for over a year when the Government escorted the United States

into the war. And to feel how my cowardice held me back from jail was a quiet little purgatory to be in. . . . felt the same rending of the spirit, I suppose, that a young man must feel who thinks it is his duty to volunteer as a soldier, but who can't muster the courage to do so. Jail, I knew, was where I belonged with my betters. To know this, and not to act upon it, was indeed penitential.

Near the end of the war, my wincing prayers for courage ended suddenly in an afternoon of mystic illumination, the date and circumstances of which I could no more forget than a married woman could forget the season of her wedding. Indescribable such experiences always are, I suppose. Something melted in the core of my nature, and fused it all together. I felt a sense of creative power in some spiritual act which I suppose I must call prayer. . . .

There was, I know, some close connection between this experience and a love which from its first beginning made moonlight of the youthful one. This one is love. It makes even my social emotions seem timid, prefatory to social emotions yet to come. Whereas the youthful one had been requited a little, requited as much perhaps as it could bear, the new one is most utterly unrequited. Perhaps all the more for that reason, it swept away fears and forebodings from my whole life. It flooded my thoughts waking and asleep with incredible delight. It played unthought of music in secret on the strings of my sedate middle-aged personality. The first few years of it made me literally live in Paradise.

This was when I kept it clean with "prayer."
I still don't know what else to call the faculty by
which we escape from self by act of will. Since
my illumination at the end of the war, I sometimes relegate myself to its place as an infinitesimal
part only of the universal life. Whenever my love
grew dim and sentimental, it plainly had resulted
from falling into thinking how to try to seem
lovable or admirable to this man. But as long
as I could throw all the my-ness of it to the winds
of heaven for the whole fraternity of life to use—
hopeless, isn't it, to try to express these things,
familiar as they become! Even I who write them
can see how incoherent my words must sound.

As long as I continued to practice this indescribable act of the heart, I lived in the extreme
sweetness of delight. But my power in this practice grew faint, and my love lost the brightness of
its rising. It has relaxed, it is a little faint. I
believe it will recover its freedom and vigor again.
But it may be that its greatest days are gone, and
will not return. . . .

My darling love had made me deeply contented
with age and death. It had made me feel profoundly, securely at home in the universe. Is it
possible that the bliss of loving one lets us each
foretaste the immeasurable bliss of loving all,
which a long transit of the heart, through lives
upon lives and deaths upon deaths, will finally
bring?

Here is where the new upsurging shattering
response of love broke down the established system
of habits and resulted in that experience, which is

responsiveness to the fullness of the event. This was the "afternoon of mystic illumination." She describes it most accurately when she says: "Something melted in the core of my nature, and fused it all together." What did she experience in this state? What was the object being experienced? She says: "I felt a sense of creative power in some spiritual act, which I suppose I must call prayer." Plainly the object of her experience was that which produced in her the religious attitude and had for her all the values of God. And out of his state of deliquescence of personality, there arose a new organization of the self, a new meaning and purpose in living. She surrendered herself to "the whole fraternity of life to use." She came to feel "profoundly, securely at home in the universe."

It will be noticed that it was not love alone that brought on this state of mind. She had been under going a most severe conflict of responses, which threw all her habits into a state of instability. It only required the additional surge of a great new emotion of love to destroy the equilibrium of habits quite completely and reduce her to a state of total aliveness, but without definite attitudes, selective attention, or fixated forms of response, and so bring her to the afternoon of mystic illumination.

Here we have that experience which we all along have described as uniquely religious. There is a reorganization of the total personality. There is a fusion of conflicting responses, a deliverance from those inhibitions, suppressions, fears and hesitancies which had hindered her living. The spontaneous, exuberant, unhindered living that resulted was the result of such deliverance. It is the universal work of religious experience when sufficiently profound. It is the supreme function of religion in human life. It is something which auto-suggestion, and that worship which is akin to auto-suggestion, that worship which comes to God with specific petitions, can never do.

But let us take another instance of the same thing, this time from a young married man, as described by himself. We can vouch for the accuracy and truthfulness of the statement.

> I had been separated from my wife and children for over a year. I felt under compulsion to continue my studies but I must also support my family, and, if possible, have them with me instead of remaining where they were, half way across the continent. I could do this only if I could get a certain kind of work that could be carried along with my studies. I made several attempts to get such work but failed. Finally I received a tentative offer, providing I could make good. The first time I came I was very nervous, being worn with much study and having lived in much isolation in the attempt to complete the work for my degree. I did not do myself justice. I came a second time and at the close of the day was told that my employers were not satisfied with my services and that I could not have the position. It was nine o'clock at night at the time. I had to ride many hours on the interurban to reach my place of residence. I shall never forget that long ride of misery. It was after two o'clock A. M.

when I got into bed. I could not sleep although I was worn with nervous strain, the day's work and disappointment.

The hours of the night were almost unendurable. Worse than the disappointment and failure to find a means of seeing again my family was the sense of my own worthlessness and futility. I felt completely beaten. It was total loss of self-confidence. It was not this last failure alone that overthrew me, but a series of experiences during the two previous years, which I had interpreted as failures. Because of these experiences I had been fighting the sense of failure and futility for some time. Now it rose up and crushed me quite completely. I had no inclination to suicide (although the thought occurred to me), because it appeared to be about the meanest thing I could do toward those who were dependent upon me. But it occurred to me that if I had only myself to consider I should certainly have gladly welcomed any means of ending the misery and futility of it all.

During the forenoon of the next day I attended to certain duties and in the afternoon returned to my room to face the facts as squarely and completely as possible and somehow find myself. I felt there was something must be fought out and settled, although I couldn't tell just exactly what it was. I suppose it was a vague sense that I must settle the problem of living my life and my relation to things in general. I felt there could be no rest for me until I settled things somehow.

I spent about four hours in my room alone. It was not exactly thinking, nor exactly praying, although at times it was one or another of these quite distinctly. Most of the time, I suppose, it was a sort of combination of these. Gradually there emerged within me a spreading sense of peace and rest. That almost unendurable pain of mind that had possessed me for twenty-four hours assuaged and passed away quite completely. I imagine it passed somewhat as a pain passes under an anæsthetic. Then I found myself filled with a strange new exuberance. I was almost laughing and crying with joy. Joy about what? I could not tell. I only knew my pain was gone, and I was full of great gladness, courage and peace. All the facts were exactly as they had been and I saw them more plainly than ever. My family was still as far away as ever and there was no visible means of getting them any closer. My failure to get the work I wanted, stood as it had before. I cannot say that I had any anticipation of how my difficulties might be overcome; I did not even have the feeling that they would be overcome. I simply knew that I was glad, and ready and fit to go ahead and do whatever I might find to do and take the consequences whatever they might be. There was no hysteria and no hallucination about it. The strong emotion of gladness gradually passed away in the course of days, but the courage, peace, readiness to meet any fortune with equanimity, and joy in living did not go away. The old anguish did not return.

In the course of time I was able to have my family with me, but we spent the winter fighting to save the life of our youngest. For six weeks his life hung in the balance and for many days we had to watch over him constantly day and night.   At one time he lay apparently dead and was only restored by artificial respiration.   Yet through all this the old feeling of dejection and failure never recurred.   Instead a deep inner feeling of calm and divine presence was with me.   In the years since then I have not kept to this high level, but I feel I have discovered the sources of infallible support in any time of need.

Now this experience would be explained by some psychologists as a case revealing the "ambivalence" of personality. According to this theory, beneath the dominant attitude there is always a suppressed attitude. When the dominant attitude wears itself out by continuous exercise, it finally sinks into the subconscious and the opposite attitude takes its place. If the dominant attitude has been that of melancholy, the succeeding attitude is that of joy. And the more extreme the melancholy the more joyous will probably be the opposite attitude that succeeds it. This mechanics of balance is the law of the mind, according to this theory; and the one attitude succeeds the other by a natural law of the mind.

Now it is certainly true that in those diseased minds which the psychiatrist studies there is always more or less extreme dissociation of opposite and conflicting attitudes. When one is dominant, the other is suppressed. And this suppression and conflict to a less degree is found in all minds, although the more healthy the mental state, the less of it there is. We do not deny that such reversal of dominant and suppressed attitudes occur. We only claim that when there is such dominance and suppression there is another way in which transformation may occur besides that of the reversal of ambivalence. In this reversal there is no genuine change of character. There is no rebirth. There is simply change of place, as it were, of two contrasting traits of the established and unchanging character. But besides this change of place, which certainly may occur, there is also such a thing as reorganization and fusion of such nature that the melancholy attitude with its characteristic impulses and conflicts, is not only suppressed, but is reorganized and fused in such a way that the total unified self emerges. This is rebirth and is a very different thing from mere substitution of one attitude for another. We believe the case mentioned is one of rebirth and not of substitution. However, until it be more adequately studied, there is room for doubt. But let us give a clear case of suppression and substitution in order to contrast it with what we have described as rebirth. Both suppression and rebirth are no doubt common occurrences in religious phenomena. But it is very important that they be not confused.

The kind of conversion which we want to contrast with rebirth might be called hypnoidal. To illustrate it we shall turn to fiction, a story in the *Atlantic Monthly* for July, 1924, called "The Devil's Instrument." While it is fiction, we believe the reader will agree with us that this is a case of fiction being truer to life than many a biographic account.

Tim Messer is finally prevailed upon to go to the meeting house with Sam, his old companion in sin, but now a new convert. The house is packed, the singing overwhelming. The preacher portrays the end of the world, when smoke begins to roll and a voice "cries out from land to sea, 'Time is no more! Time is no more! . . . Where will you be! Where will you be when that last trumpet sounds. Sinner man, yea, you, Tim Messer, where will you be. You will cry for mercy and there will be no mercy.' "

Tim felt his flesh freezing as the terrible picture grew. At last the preacher ended his description of Tim's likely end by imitating the wails and screams that would rise to heaven when he lay in outer darkness. His yells and bellowings swept a score of listeners to their knees. Several little children were already writhing on the hay, and men were crying in different parts of the house. Here and there old women crouched at their seats agonizing over Tim's soul. As if in a dream, he heard them mentioning his name.

Then Brother Baxter (the preacher) turned to the table and drank deeply from the pitcher of water. He came back to Tim and pleaded with him in a gentle voice.

At the sweetness of his words, Tim felt the hot tears go coursing down his leathery cheeks.

Finally the preacher described how the devil

would come to get Tim's soul as he lay on his death-bed.

> He'll come in easy, oh, so easy, his tail making a grisly sliding sound on the floor. Then! [he roared, seizing the terrified Tim by the shoulders,] He'll jump onto the bed and carry yer soul screaming to everlasting hell!

In the story, which we cannot here quote in full, it is easy to trace the process of hypnosis, the dissociation of one set of impulses from the rest of the personality, the fulfillment of these, and the later conflict between these newly acquired habits and the old system of habits which still constituted the major portion of his character. There had been no breaking down and deliquescence of established habits with a reorganization of the total personality into a new and different system of habits. There was no rebirth. There was only dissociation and suppression. Certain impulses, which have been more or less dormant from childhood, or which may have been gradually maturing during the days preceding, were so stimulated at the meeting, and so dissociated from all counter impulses, that they were brought into control of conscious conduct, and for a time were able to hold all other habits in suppression. The revival meeting and the preacher were splendidly adapted to dissociate these impulses from the inhibitions of Tim's established character, but they were not adapted to bring him into that experience which would transform his characer.

In hypnosis and auto-suggestion we have disso-

ciaion and suppression; in the profounder religious
experience we have deliquescence of old habits and
out of the resulting free play of impulses there
arises a new, harmonious, unified self.

William James notes these two ways in which
the individual may be changed. He says in *Varieties
of Religious Experience:*[1]

> There are only two ways in which it is possible
> to get rid of anger, worry, fear, despair, or other
> undesirable affections. One is that an opposite
> affection would overpoweringly break over us, and
> the other is by getting so exhausted with the
> struggle that we have to stop—so we drop down,
> give up, and don't care any longer. Our emotional
> brain-centers strike work, and we lapse into a
> temporary apathy. Now there is documentary
> proof that this state of temporary exhaustion not
> infrequently forms part of the conversion crisis.
> So long as the egoistic worry of the sick soul
> guards the door, the expansive confidence of the
> soul of faith gains no presence. But let the for-
> mer faint away, even but for a moment, and the
> latter can profit by the opportunity, and, having
> once acquired possession, may retain it. Carlyle's
> Teufelsdrockh passes from the everlasting No to
> the everlasting Yes through a "Center of In-
> difference."

James' comment would seem to indicate there
are two souls or systems of response, one the sick
soul and the other the soul of faith, and that the
soul of faith existed somehow down in the subliminal regions of the mind, suppressed and held

---

[1] *loc. cit.* p. 212.

there by the dominance of the sick soul, and when the latter became exhausted this soul of faith could spring forth from the subconscious and take possession of the individual.

We believe this multiplication of souls and selves has been altogether overdone. We shall criticize this Freudian theory more fully in chapter XI. The "Center of Indifference" may issue in the substitution of one old self for another. But it may also issue in the fusion of old selves and the emergence of a new self. This latter eventuality has not been sufficiently recognized, even by James.

We cannot have continuously the mystic experience of God. The practical, biological and social requirements of living forbid it. To live we must have an efficiently organized system of habits. But such habits necessitate a very narrowly selective attention and a very rigid constraint of impulses. Our hope of growing into an ever more abundant life lies not in discarding all such systems with their limitations and constraints, their routine and cramping confinement, their deadening of spontaneity and originality. No, we must continuously harness ourselves with such habits. Without them not only would all our efforts be wholly futile, but we would cease to have any definiteness of character, of purpose, of thought and perception, of individuality. We would not even have that originality and spontaneity which the system of habits seem to destroy. Without such habits there are only two alternatives before us. One is the stupor of profound sleep or total loss of consciousness, the mere vegetative existence, due to the reduction of all response to the very fewest impulses. The other is the melting glow and exhilaration of the mystic experience without definite thought or purpose, due to the increase of response through the multiplication of free impulses so numerous and diverse as to prevent the organization and fulfillment of any of them. Between these two extremes lies the ordinary human way of living with its established and indispensable habits.

How, then, can we hope to grow into an ever more abundant life? Only by a method of alternation between the mystic experience and the life of organized habit. To grow into an ever more abundant life we must discard the old self from time to time and take on a new self. We must cast aside the old system of habits and enter into a new system. We must break the old, ever hardening shell of habit and for a brief time live without a shell, palpitating and throbbing with the flood of experience flowing over us. But then, in order to deal with experience at all, we must develop another shell of habits, develop and perfect it for a time until we have grown to its limit. Then we must again break through and expose our sensitivity to the full stream of experience, and out of the innumerable new impulses thus aroused form another and better system of habits. Only so can we be lifted out of the ruts. But forthwith we must make new ruts if we are to travel at all. These new ruts should enable us to travel further than the old ones. We must escape from ourselves from time to time, in the experience of God; but we must come back to ourselves if we are to have any selves at all. But the self to which we come back should be a richer, greater self, because of this experience of God. Only in this way can the panorama widen forever about us. Only thus can the spirit of the little child, which grows eternally, retain the freshness and vitality which God has given it. To experience God is to turn and become as a little child. Hypnosis and auto-suggestion are the fine technique of old age; but experience of God is the rebirth of eternal youth.

CHAPTER IX

CHRISTIANITY AND SCIENTIFIC MORALITY

Much has been said of late of the need of scientific method in morals. Conscience and moral endeavor must be shaped by the deliverances of science. Scientific technique must be applied to set forth the essential needs of human living; to distinguish between genuine and illusory wants; to ascertain what wants can be harmonized with one another and what cannot, and to devise a method for achieving this harmony. What sort of environment, physical and social, do the essential interests of human living require? And what form and method must these interests assume in order to attain maximum fulfillment? These are questions which science alone can adequately answer; and they must be answered if we are to solve our moral problems. Unquestionably scientific method must be put into our morality and we can never have too much of it.

But the same fatality which overcomes scientific method elsewhere, affects it when applied to the moral life. Scientific method is indispensable, but it is not self-sufficient. Scientific method can bring harmony and fulfillment to our wants but it cannot to any great degree create in us a new and richer system of wants such as we saw occurs in rebirth when God is experienced. It can bring maximum efficiency to our old self; but it cannot give us a new self. The more finely coördinated becomes our social system, the more perfectly adjusted to our needs becomes our physical environment, under the prescriptions of science, the more firmly fixed and unchangeable become our interests. It is not that we lack the power to do what we will, but that we lack the vision to seek any other ends than those prescribed by this most efficient system of life which scientific method has established. The deepest craving of human nature is for growth, for ever increasing abundance. But scientific method cuts off growth. Such a finely adjusted life leaves nothing to develop save the perfected system. This may be made more efficient and we may seek change and excitement by going round and round and round, faster and faster and faster. But we cannot escape the system. Why? Cannot science see this need also and provide for growth? Provide for escape from the system? No, we have already seen the reason why it cannot. It certainly widens the range of our interests in space and time. But it cannot create in us new interests. We do not now refer to the great creative scientist, but to the effect of scientific technique upon the average run of scientists and upon the mass of mankind.

The horror of this scientifically efficient life is beginning to creep over our civilization. It is beginning to appear in fiction and essay. The novels of Sinclair Lewis portray it. It is reiterated in Mumford's *Story of Utopias*. L. P. Jacks does not weary of declaring it. The restlessness and discontent of great masses of people reveal it. There is a wave of discouragement and disillusionment passing over the world that is not altogether due to the after effects of the war. It is due to discernment, only half conscious, that the much-lauded progress brought on by scientific method, while undoubtedly progress in the sense of rendering us more efficient in satisfying our wants, fails to create in us an ever richer and fuller system of wants. It fails to lead us into an ever wider and fuller vision. This is what Chrisian mysticism must do. This is the chief contribution of Christianity to morals. It counteracts the inherent fatality involved in scientific method.

A scientific morality can promote the development of such a system of habits in the individual. such a system of social order and such adjustment of physical conditions, as to give tolerable expression to all recognized and worthy ideals. But it cannot give rise to radically new ideals. It cuts off the growing edge of life. With the help of the sociologist on the one hand and the psychiatrist and biologist on the other, the exact sciences can remedy to a large degree the evil of thwarted and suppressed impulses—save only that undefined human craving for an ever fuller life.

We have already seen in the previous chapter how Christianity meets this deepest need of all and supplements scientific method.   But let us go more deeply into this matter to show how scientific method in morals needs the support of Christianity.

THE MORAL DYNAMIC OF CHRISTIAN WORSHIP

Much has been said of the influx of power from the subconscious brought about through worship. The question has been asked whether this influx came from some source external to the individual, such as God, or whether it was the release of energy stored in the organism, and hence having its source internal to the individual.   But such a question, we believe, shows a total misapprehension of the whole situation and of that basic relation called stimulus and response.   As though the source of the energy must be either external or internal!   Of course it is both.   The energy of the total organism can be released only when the proper stimulus is applied. Is, then, the source of the energy the stimulus or the organism?   It is both.   The energy was unavailable for work until the stimulus was applied.   It is the stimulus which renders the energy dynamic.   It is through worship that one exposes himself to that stimulus which is capable of awakening all the power of the individual and releasing the stored energy of the organism for work.   Besides, organic activity of any kind always involves activity of the environment just as much as activity of the organism.   For example the lungs cannot breathe apart from the air.   The air does the breathing as much as the lungs.   Breathing is a total complex process of interaction between organism and environment, in which the environment is just as active as the organism. What is true of breathing is likewise true of anything else one may do. It is true of walking and thinking, desiring and willing, dreaming and aspiring. None of these is done by the body alone nor can it be found in the body any more than walking can be found in the foot. Always the environment plays as large a part as the organism in the complex process of interaction. Whence, then, comes this energy, this influx of power? Plainly it does not come "from within" any more than "from without."

The problem of how to release maximum energy in human living is largely a problem of how to bring all stimuli to bear upon the organism so as to awaken its full capacity for response, but in such a way that the responses shall not obstruct one another.

Worship does this. In mystic worship the attention is no longer selective. One attends in a diffusive way to the total mass of stimulation which is playing upon the organism and in it. But one does not attend to these as so many diverse stimuli; rather he attends to them as a single total mass of stimulation. They do not awaken in him divergent impulses, because his attention is so diffusive and there is no specific response. The mass of stimulation quickens his powers in a uniform pervasive manner. Something analogous to it is found in aesthetic contemplation. His whole system is toned up. The habits and impulses of which one is capable are aroused simultaneously and produce within one a labile interplay of a thronging mass of impulses which merge into a single undefined quickening of the total personality. The resulting state is one of emotional glow, of calmness and poise, a sense of command over all one's powers and a readiness to face whatever the world may offer. There is no specific response, but a readiness to respond with utmost energy and self-control in whatever way may be required. For some, it would seem, a sort of ecstacy is added.

It should be noted that what we have described is a generation of energy, but without any direction given to it. The energy is not put to work. If this is as far as the matter goes it does not accomplish much. It gives one a few minutes, or it may be a few hours, of peace or ecstacy, and then passes away. It is a sort of orgy in which one indulges. It is a state that comes and goes and leaves nothing behind except that toning up of the whole organism and that relief from tension and strain. There is a peace and power that continues; but so far as concerns any constructive remaking of self and the world, there is little or none that results. In order to have these constructive results there must be something else. Christianity supplies this something else.

What we have been describing is the dynamic of mystic worship in general, but not necessarily Christian. Unchristianized mysticism is a dangerous thing, just as maximum energy undirected is always dangerous. The energy made available through such mysticism may overflow into all sorts of excesses.   The peace and power brought on by it may operate in a destructive or unhealthy manner.

THE DYNAMIC OF CHRISTIAN WORSHIP

The Christian does not merely submit himself to the mass of stimulation reached through mysticism.   He does more;   he endeavors to interpret this stimulation.   He comes to it with a well-defined belief and purpose which together are called faith. His faith is that this total mass of stimulation is working upon him and his world to reconstruct both.   He has some idea concerning what this stimulation would make out of him and his world.   He has the faith that he is conversing with a Will and that this Will is working to refashion him into the likeness of Jesus Christ;   and, further, that it is working to reconstruct his world into the Kingdom of God as depicted in the words of Jesus.

When one enters into mystic worship with such a faith as this, it becomes very different from elementary mysticism.   It produces far different results.   Such worship becomes the most effective influence for actually reshaping the individual into the likeness of Jesus and of remaking his world, in so far as his worship can reach, into the likeness of the Kingdom of God.   Such worship leaves after-effects, not only of peace and power, but of peace and power directed into new constructive channels. Such worship is a reconstructive force working through the individual.

We have spoken of the danger of unchristianized mysticism. There is also danger in the process of Christianizing mysticism. The danger is that mysticism may be driven out altogether. Let us explain.

When mysticism is Christianized the individual presents himself to the total mass of stimulation and responds with all his impulses, but under the control of a certain belief concerning the character of this stimulation and concerning the form which his own response should assume. He believes that this stimulation is the pervasive will of God working through every organic process of his body and through every hidden recess of his mind. He believes, furthermore, that this will of God found most complete expression and embodiment in the historic person of Jesus Christ; and that this Will, which so completely fulfilled itself in Jesus, is striving to fulfill itself through his own person. He believes, still further, that, despite the fact his own activities diverge so far from the will of God, the will of God can still retain its own unchanging purpose while merging so intimately in all his own activities, because it is a will of suffering love. Thus the Christian mystic's worship is a way by which he yields himself, like copper wire to the electric current, to this Will which so persistently and pervasively works upon him to reshape him and surcharge him into the likeness of Jesus Christ.

So it is that the Christian mystic comes to his worship with a definite belief and purpose. But mysticism consists precisely in putting aside all preconceptions which narrow the range of sensitivity, exclude from attention many stimuli, and suppress many responses. Mysticism consists precisely in putting aside all definite purpose, all special interest, and yielding oneself unresistingly and without bias to the play of all the stimuli that reach the organism. One must do this in order to avoid that narrow exclusiveness and selectiveness of attention by which ordinary life keeps out this mass of stimuli and suppresses all those responses save a few. Thus there is a conflict between pure mysticism and any kind of definite belief and purpose such as the Christian must have. The Christian brings a belief that is likely to select and sift the stimuli; and a purpose that is likely to distort and suppress the responses. Consequently the Christianizing of mysticism often drives out mysticism altogether. In other cases the mysticism has driven out the definite belief and purpose which is Christianity. So it is that there has often been war between Christianity and mysticism. Some Christian leaders and institutions today are suspicious, some even hostile, to all mysticism.

But traditional Chrisianity, on the one hand, and mysticism on the other, must be reconciled if there is to be any Christian acquaintance with God. If the two cannot be united traditional Christianity ceases to be a religion at all and becomes only a mass of traditional ideas and standards of conduct enforced by certain institutions.

But the two can be united and to some degree always are wherever the Christian religion, (as distinguished from the mere tradition of Christianity) is found. Their union requires that the purpose which the Christian brings to his worship be not narrowly fixed by tradition but be held subject to modification by the mystic experience. Their union requires that the belief which the Christian brings be progressively developed under the stimulation of the mystic experience. The belief must not persist rigidly in the form in which the individual first accepted it. It must not narrow the experience down to its own measure, but must be enlarged and reshaped to fit the experience. There must be constant interaction between traditional belief and mystic experience, and the experience interpreted and its energy directed by the belief. The Christian faith (which is the combination of belief and purpose above described), as handed down by tradition, must be a device by which all the impulses of the individual, quickened to action through and through by the pervasive stimulation of experiencing God, may be organized and directed without suppression or distortion of any impulse, and without impairing the group life of individuals with one another. How to catch the personality in the full swing of mystic stimulation and direct it into constructive endeavor and appreciation without loss of energy or vision, that is the service of Christian worship to morality. To do this the Christian must have definite ideas about God and the world; and he must have definite purposes concerning right and wrong. But these ideas and purposes must be capable of much growth and transformation. The faith the Christian brings to his worship must not select and exclude; it must only direct and instruct the total personality as he responds to the immediate presence of God. For there is no stimulus which so completely awakens the whole personality as the immediate presence of God; and there is no time when the person so much needs instruction and direction.

The whole significance of Christian doctrine, and the chief mission of the Christian church, should be to provide the individual with an adequate belief and purpose for his worship. It must so equip him with a faith that he can reap all the benefits of a Christianized mysticism while avoiding two opposite evils—on the one hand, the dangers and excesses of an unchristianized mysticism, and on the other hand, the suppressions and perversions of a narrow, rigid tradition. To equip the individual with a purpose which directs, but does not suppress, any of the impulses aroused in mystic worship; and with an idea which interprets, but does not exclude, any of the stimuli unveiled in mystic worship—this is the task of the Christian church. This is a tremendous task. Too often the church has not seen that such provision for worship is the chief thing it has to do. Too often it has not recognized this to be any part of its work. Too often it has provided a doctrine and a ceremony which dried up the fountain of mysticism and thus deprived the individual and society of that beneficent, constructive dynamic which resides in Christian worship, and which is the greatest power known to man. In such worship there is power

which can make the individual over into the likeness of Jesus of Nazareth and a power which can
reshape society after the fashion of the Kingdom
proclaimed by Him.   But this power is unavailable
until the individual is equipped with plastic and
otherwise adequate ideas and moral purposes enabling him to make Christian acquaintance with God
through mystic worship.   When the church puts
anything else before this task of bringing men to
acquaintance with God and so releasing the great
constructive powers for individual upbuilding and
social service, it is recreant to its duty.

PRAYER AND ITS ANSWER

The state of pervasive stimulation attained in
mystic worship is one in which the psychological
and physiological processes of the organism are
most plastic and readily moulded into the form of
any new or larger purpose.   Hence if, in this state
of worship, one prays for health or poise, for sympathy or understanding, for skill or mastery, for
anything which has to do with upbuilding of the
self, one can have it;   for in this state the subtle
mechanisms of the organism are most completely at
the command of the individual.   In this state one
can most adequately develop these habits—whether
mental or physiological or both—by which
achievement and appreciation, adaptability and
comprehension, can reach their maximum.   This
control and redirection of one's energies is the answer to prayer so far as it has to do with one's own
upbuilding.   And one's own upbuilding should be

one's first concern.   Draw first the beam out of your own eye before you commence to meddle with your brother's affairs.

Here a question might be raised. Our wording might be interpreted to indicate that prayer is a means by which the individual attains control of himself and hence is prayer to himself and answered by himself.   But this is precisely that misunderstanding to which we referred at the beginning of the previous section, arising out of the confusion of stimulus and response.   To see more clearly that it is God who works in answered prayer, let us note a slight distinction between worship and prayer.

Prayer should ordinarily come after worship. To worship means to become wholly attentive to God, *i.e.* to subject oneself to that total mass of stimulation which is playing upon one all the time but to which one is not responsive save in worship. Then prayer is what normally follows in Christian worship.   Prayer is that purpose which becomes dominant in this state of worship.   This purpose, in normal healthy prayer, is the product of two factors; (1) the persistent desires and past experiences of the individual; (2) reorganization and unification of these by reason of their simultaneous stimulation occurring in the mystic experience. The dominant purpose thus arising, in which the whole personality is absorbed and harmonized, is the work of God inasmuch as the stimulating presence of God has brought it into existence.   It is by reason of the organization of all the mechanisms of the organism under this dominant purpose, that the marvelous results of prayer are produced in the individual. It is plainly the work of that total mass of stimulation which reaches him through the mystic experience. And this is precisely the work of God.

But one does not always pray for a remoulding of the self or for anything which can be affected by the remoulding of self. One may pray for matters which seem to lie wholly outside the sphere of one's influence. Can prayer modify such remote objects and events? Here we cannot be dogmatic. Science has scarcely begun to fathom all the different processes at work in the world. Many an established "law" of science will have to be reconstructed as new facts and theories are brought into the realm of scientific knowledge. It is quite possible that when the personality is completely awakened, as in worship, there are tensions, vibrations, influences, call them what you will, which come into play, and which produce results which could not otherwise be produced, even when those results seem to lie completely beyond the control of ordinary human effort or of this particular individual's effort.

This hypothesis concerning prayer will be wholly unsatisfactory to some. They will say: In prayer it is God who must produce the result and not the fully awakened personality with its vibrations, tensions, influences, etc. Of course it is God who produces the result. This we have already explained. It is God because it is He who through worship awakens the total personality and so gives rise to these novel forces. It is just as truly God producing the results through prayer as it would be if one imagined God up in the sky listening to the prayer and then sending an angel down to a designated locality to do what was required. It would be God operating in both cases, but in the hypothesis we suggest the operation would not conflict with a conceivable science. Both descriptions of prayer, that of God and angel, and that of tensions and influences, may be equally correct and equally false, inasmuch as both are theories by which we endeavor to designate processes which we believe are going on but which we cannot describe. Until we can describe them more accurately we must content ourselves with such theories.

SOCIAL SOLIDARITY THROUGH WORSHIP

There are two ways in which people may reach agreement. The first includes the second, but the second can stand by itself without the first. The first is agreement through sharing the same mass of sensuous experience. When my brother and I have seen and felt the grateful shade of the same tree, listened to the rustle of its leaves and viewed the landscape from its base, we can agree concerning many things pertaining to that tree. We have shared a common experience. This is the sort of agreement which may be called community or brotherhood. This is the agreement which makes the whole world kin because it is a common experience of nature.

But there is another kind of agreement which is purely logical. It is agreement due to the compulsion of logical implication. It is agreement concerning abstract propositions, such as: Two times two equals four; the three angles of a triangle are equal to two right angles; if *a* is to the right of *b* and *b* is to the right of *c*, then *a* is to the right of *c*. Such logical agreement does not of itself yield brotherhood. It does not constitute community. It does not bind hearts together. No doubt it is required in order to have the community *of a* common mass of sensuous experience, but by itself alone it does not bring about community of purpose and good fellowship. There need be no sympathy, no love, no joining of common human purposes in such agreement. We mention this kind of agreement only in order to get it out of the way and show that it does not yield social solidarity.

Our proposition is that all community of purpose, all depth of sympathy, in a word, all social solidarity, is based on that agreement which arises out of sharing a common mass of experience. The more such common experience is shared, providing there is sufficient agreement in logical definition and scientific description to enable the individuals to recognize the community of their experience, the more social solidarity there is.

The problem is how to attain more of this recognized community of experience which yields brotherhood. The experience gathered in ordinary conduct of life is very defective in this respect because in ordinary conduct we select from out the mass of experience only those features which meet the demands of the peculiar practical and theoretical interest which dominates us as individuals or as representatives of a certain group. Other individuals, and representatives of other groups, select different features. Hence we cannot agree; we cannot attain community. Even when we do select the same features each gives it a caste different from the others because of his own particular bias. Only by attending to the total mass of experience, unsifted, and undistorted, can we attain this fuller community.

This, we have seen, is precisely what worship does. In worship we cast aside our special bias and selective interest. Through worship deep and dear community is most completely attained because in worship we share the common mass of experience poured out on all men. In worship we attain that brotherhood which binds together the hearts of men.

There is something else which promotes this same good, although to a less degree than worship. It is art. Art takes from us the bias of practical and theoretical and personal interest. Aesthetic appreciation is disinterested. In so far as many enjoy the same works of art they share the same features of sensuous experience. Not only do they share these features in the works of art, but through these works of art they are trained to take note of these features in the natural world round about them all the time. A color, a contour, a sound, a movement, will be consciously noted by them because they have had it impressed upon them in works of art. It enters into that mass of experience which constitutes the background of consciousness. This common background, shared by many, gives community. They who sing together the same songs, dance the folk dances, inherit the same myths, stories and poems, and view the same paintings, have a social solidarity which cannot be attained by those who lack such a common heritage. With such common heritage of art must be classed also certain conventions of speech, of intonation, of good manners and standards of polite society in dress and gesture. All these are forms of art and direct the selective attention of all to the same features in the world of nature so that they share a common mass of experience.

But worship goes much deeper than art. It does not merely guide the selective attention to common features that can be aesthetically appreciated, but it goes on to that greater mass of common features which can be appreciated only mystically. The Greeks were bound together by art, the Jews by worship. History shows which is the stronger tie.

MORAL VISION THROUGH WORSHIP

The greatest difficulty in moral conduct is to keep all the relevant facts in view. If we could do this, nine-tenths of our moral problems would be solved. But we cannot keep all the facts in view. Passion and prejudice distort the vision and becloud the issue. Certain facts stand out gleaming white, like peaks caught by the setting sun, while all the rest lie hidden in the shadow. Passion and prejudice illumine those peaks which lure us and leave the others in darkness. By passion we mean the dominant impulse which has been aroused by the novel situation. By prejudice we mean the habits, due to past training, which are so rigid they will not yield to the pressure of new and more complex situations. The man who is controlled by the routine of habit is not ordinarily swayed by passion; and the man swept with passion is not commonly the creature of rigid habits. But each throws certain facts into bold relief while other equally pertinent matters are ignored. The problem of moral conduct is how to see all facts in the clear full light of reason instead of those few isolated facts in the slant light of passion and prejudice.

Now worship, as we have seen, is that attitude in which hidden things come to light, the selections and sifting and exclusions of attention are more or less removed so that all things are viewed in more of a uniform light. Something approximating noonday, or, if one prefers, a pervasive twilight, is substituted for the flashes and streamings of passion. It is in worship, then, that the great moral problems are best solved. He who cannot worship, or who cannot retire to a quiet place and "collect himself," cannot deliver himself from the clutch of passion and prejudice sufficiently to react to all the pertinent facts. Hence such a one cannot live a very moral life. That is to say, he cannot master the technique of abundant living. Such a one may be most highly respectable. The man whose narrow prejudices and passions happen to fit into the social order of his time may live a most faultless life from the standpoint of the administrator who notes only flagrant abuses of the mores. But he does not live an abundant life, and morality is the technique of the most abundant life which the mores permit. Within these limitations of the mores there is a minimum and a maximum. If narrow prejudices have been shaped by careful training, they will keep one at the minimum. But only high morality will bring one to the maximum. And high morality requires worship, for the reason stated.

MORAL CREATIVITY OF WORSHIP

A religion which is devoted exclusively to the promotion of human welfare, as current opinion has defined human welfare, becomes a servant to the prevailing arts and sciences. Its God becomes identified with the values which the people of that time and place have come to recognize. Religion then becomes the great sustainer of the customs and traditions, the arts and sciences, that constitute the civilization of the time. Such a religion cannot be a transformer or revolutionizer. Such religion cannot point men to other values on beyond those goods which they seek in their everyday lives. Such a religion is not a master of the prevailing civilization, but a servant. Such a religion cannot usher in a new day, it can only help to perpetuate an old day. The voice of God, in such a religion, is merely the echo of what the custom and the sciences declare. When the statesman or scientist says that such and such a matter is good, it becomes the business of such a religion to respond: Thus saith the Lord, and so herd the masses into conformity with the dictates of scientist and statesman. It was such a religion that the Roman Caesars tried to bring to perfection in demanding that all men worship Caesar as God. The Romans were intensely practical people and their religion was a practical religion. No mystical nonsense for them.

Early Christianity was not a practical religion in this sense. It did not sustain the values which were socially recognized by the civilization that prevailed. It did not promote social solidarity under the Roman government. That was precisely the reason why the Roman government persecuted the early Christians. Early Christianity was a revolutionary religion. It was a ferment that disintegrated the prevailing social order in the interest of a totally different manner of life.

Religion at its best is always creative. To be creative means to run counter, in some respects, to the established practices of the day and so to hinder, instead of promote, the recognized practical and social requirements. To be creative means to introduce new values beyond those which men have heretofore recognized and to devise new forms of conduct different from those which the established social order and the prevailing arts and sciences prescribe. To be creative means to be sovereign, not merely servant, to the prevailing civilization. It means to lead, not merely to follow.

The creation of new and more satisfactory ways of living is one of the noblest of all the arts. There is more creativity in it than in poetry, painting or music. It requires more courage, more insight and more artistry than any of the fine arts. Indeed there are grounds for saying that the fine arts are the refuge to which creative natures flee who have not the courage and mastery over men and conditions sufficient to exercise their creative talents in that more direct moulding of human life which is found in devising new and better habits of dealing with environment. The great originator of moral forms of conduct is the artist who has not been forced to shrink back to the handling of such unresisting materials as paint and sound and stone, but has gone straight to the hearts of men and wrought his new creations directly into their flesh and mental habits. He is the prophet who has had the religious experience which dissolves old habits and views and makes possible the rise of new outlooks, purposes, values. He does not merely echo the pronouncements of secular leaders, but gives original pronouncements of his own. These tremendous transformers of human ways of living have come, after spending forty days and nights in worship, saying: Ye have heard of old time, an eye for an eye and a tooth for a tooth, but I say unto you, love your enemies. They have said that when you are struck upon one cheek, turn the other. They have said that the Lord sends his rain upon the just and the unjust, and have exhorted us to be perfect as our heavenly Father is perfect in this respect. Such revolutionists as these have been mystics, cultivating the presence of God for God's sake. Such were Jesus, Buddha, Confucius, Moses and Mohammed.

It would seem from these facts that worship is one of the sources out of which new creations in the art of living arise. It is in worship that new paths open up; worship is the only suitable preparation for the greatest creative artistry in all the world, the art of reshaping the total vital process of living. This is morality. Through worship we reach a vantage point outside the civilization of any particular time in the sense that we can survey it, criticize it, find it wanting and reconstruct it. Worship lifts us beyond civilization. Worship is uncivilized and can never be wholly civilized. It is wild and untamed. It comes in to destroy civilizations and to make them over again. It is from worship that the great religious leaders have come back to destroy that interlocking congeries of customs, which determine the technique of living (moral standards) of any time and people, and to create a new technique, which means a new morality. They return from the wilderness with the vision of a new world in their eyes.

Here, we believe, is the correct adjustment between morals and worship. Religion should promote the welfare of man but it should do so not merely by conserving the socially recognized values and inspiring men to achieve or to conform as sec-

ular leaders may determine. It may very properly
do this, to be sure. But it must do more than this
if religion is to preserve itself and not be sucked
down into the stream of tradition and become
merely an established institution like any other
system of customs. Its chief social function is not
to support the established order, or secular leader,
although it may well do this, but to take the place
of leadership itself from time to time. And this it
can do only when it preserves worship as something independent and insubordinate to the recognized goods of life, because of its devotion to the
unrecognized goods. Worship cannot be wholly
subordinated to the promotion of human welfare,
if by welfare we mean the attainment of those
goods which we have thus far been able to discern.
Worship must serve to bring to our ken goods thus
far undiscerned and unacknowledged by our civilization. If it is not guarded in this function, but
given over wholly to inspiring human conduct in
the quest of established goods, the chief fountain
head of increasing life for man will be stopped.

The more scientific our morality becomes, the
more it requires worship to provide for growth.
The more strenuously we strive to attain the
highest morality thus far discerned, the more incapable we become of discovering any other forms
of conduct which may yield a richer life, and hence
constitute a higher morality. The more we con-
centrate our attention and specialize our effort
upon the attainment of certain definite goods, the
more completely do we shut out from our field of

vision any other goods. The more finely adapted, *i.e.*, scientifically constructed our social organization becomes for the maintenance and acquisition of certain values, the more strictly does it confine our range of interest, and hence eliminate those blurred, undefined borders of interest which constitute the growing edge of life. These features of scientific moral efficiency (a) strenuosity, (b) concentration and specialization of attention and effort, and (c) the fine adaptation of social organization to certain well-defined functions, all three tend to confine human living to certain limits and to bring growth to an end. It is only as we can break through the charmed circle of such a system, and bring to light new and different goods to seek, that we can have growth.

What is more, when growth is thus cut off and the goods we seek are so strictly defined, they lose their charm in our eyes. The greatest lure of any object of endeavor is that larger, more or less undefined good, to which it may lead. But when the scope of endeavor and aspiration is so clearly defined and limited, this great lure vanishes. Life becomes stale. Vanishes the glamour and romance, which is simply our sense of the great unattained and undefined beyond. And with this vanishing of the undefined possibilities of growth, human life ceases to satisfy.

This is the reason why moral creativity, or the growing edge of life, must be guarded above all. This is the reason why Christian worship is necessary to supplement scientific morality.

How does worship keep open the door toward the unattained and undiscerned goods? The psychology of worship, as we have previously described it, plainly reveals why worship does this above all else. Our sense organs are capable of revealing to us far more than we have thus far learned to interpret, recognize, or appreciate in any way. Besides the external sense organs, which themselves can deliver to our apperception immeasurably more than we have learned how to appreciate, there are the internal organs, glands, muscles, etc., all of which signify something with regard to our total environment. To worship means to surrender our attention to this mass of experience from all these sources. Out of experience must be wrought whatever further goods may ever be attained by us. Worship provides for growth because it brings to awareness such masses of experience.

In worship we go back, as it were, to the beginning, before the paths of selection and interpretation diverge. If we are to beat new paths through the jungle of experience to the winning of new goods, we best can do it by thus returning to this starting point of unsifted and undefined sensuous experience. This is where the infant begins his acquisition of the arts and sciences of his day; and the goods which civilization has to offer him must reach him through these sense data. The significance he is able to work out of these sense data, and through them, is the measure of what he will be able to acquire. And this mass of sensuous experience is where the human race began its historic development of all the arts and sciences.

But when the mystic in worship returns to this starting point, he is not an infant or a savage. We have compared him to these only because, like them, he deals with uninterpreted experience. But he differs from them on two points. (1) The infant and savage are not necessarily aware of any wide fullness of immediate experience, as the mystic is. (2) The infant and savage face whatever data of sense enter their awareness, with little notion of any great undefined significance in them. They know nothing of those larger goods which may be wrought out of these data and the fullness of fact which they must signify, however undiscovered that significance may be. But the mystic does know there is such wealth of significance in immediate experience, because he has returned to this veil of immediacy from out of the system of arts and sciences which constitute the culture of his day. He has been beyond the veil, and has seen the great world as his age has been able to depict it. Now he comes back again behind the misty undefined veil of sense. What lies beyond this veil? What does it signify? Well, it signifies all that the arts and sciences of his age have been able truly to discover. But is that all it signifies? No, indeed. No scientist would say that science had yet discovered all that the data of sense signify. No artist would say that art had brought all possibilities of beauty to light. No moralist would say that all noblest programs of action have been achieved. No sociologist or statesman or group of lovers would say that all the joys and achievements of human association had yet been reached. But above all the mystic knows, when he returns from that vision of the world which the culture of his day presents, back again to that mass of undefined experience out of which this vision has been developed, that this vision does not reveal more than the tiniest fraction of the total significance of this which he now immediately experiences. Vast regions of unexplored significance are in these data, or in this total datum, which he is now experiencing. God is there; and what enters into the life of God is there. He may feel there is this total significance in his mass of experience, even though he can in no wise define it and has no descriptive knowledge of it.

So it is that the mystic stands at the point from which all new paths must be broken. His experience is that out of which all new creations must be brought forth. What is the vast and total significance of his experience he does not know. But the civilization of his day, and above all, his religious heritage, has taught him that it has such significance. That is the reason he is a worshipper while the infant is not. That is the reason it produces in him an attitude of anticipation, of wonder, and of awe.

But these masses of experience which the mystic brings to light, can yield up their meaning only as they are interpreted. As long as they are not interpreted, as long as they merely flood consciousness with a sense of vast undefined meaning and tone up the organism with their pervasive stimulation, they are a form of luxury. They are forms of dissipation in which the mystic may revel but which are of no value to anyone else and of no value to him after the mystic hour has passed. These masses of experience must be made to yield up their significance.

Now the interpretation of experience, making it yield up its significance, is the work of. science The prophetic mystic may guess at the significance. He may sometimes strike wonderfully near the truth with his guesses. He may have flashes of insight which are sometimes amazing. But his guesses will sometimes be just as amazingly fantastic. There is no method of testing truth and distinguishing it from error save scientific method.

Science needs religion to provide it with the raw material of fresh experience and fertility and plasticity of imagination. Scientific method cannot lead on continuously to ever more abundant life unless it is supplemented by religion.

PART III

THE NATURE AND FUNCTION OF RELIGION

CHAPTER X

RELIGION AND IDEALS

Can we identify religion with devotion to ideals? There is to-day a widespread tendency to do this. Prominent in a public library stands the motto: "To be religious is to pursue the highest ideals." Sometimes the word value is substituted for the word ideal; but the word value is no clearer in its significance than ideal and there is just as much difference of opinion concerning its exact meaning. Is that which we experience when we have distinctively religious experience an ideal? Is God or Christ or whatever the religious person may consider the chief object of his concern, preëminently an ideal? Is the chief function of religion to clarify, enforce and make alluring certain ideals—"the conservation of socially recognized values"—for instance? Is religion preëminently a device for glorifying social coöperation and arousing utmost devotion to those goals of endeavor which society holds to be highest?

Or is the function of religious experience relative to ideals creative rather than conservative? Is it the reconstruction of ideals rather than the enhancement and enforcement of established and recognized ideals? Is it "the revaluation of values" rather than the buttressing of values hitherto recognized? Is it the *experience of that from which new ideals may be derived* rather than the experience of ideals themselves?

It is the second of these two suggested positions that we hold. We are very sure that the greatest obstacle in the way of individual growth and social progress is the ideal which dominates the individual or the group. The greatest instrument of achievement and improvement is the ideal, and therefore our constant failures, miseries, and wickedness are precisely due to the inadequacy of our highest ideals. Our ideals have in them all the error, all the impracticability, all the perversity and confusion that human beings who are themselves erring, impracticable, perverse and confused, can put into them. Our ideals are no doubt the best we have in the way of our constructions. But the best we have is pitifully inadequate. Our hope and full assurance is that we have in religion that by which constantly to reconstruct these ideals, casting them aside ever again for better ones. Our hope is that we can improve our ideals. If we could not be saved from our own ideals, we would be lost indeed.

If religion were identical with devotion to our highest ideals it plainly could not deliver us from the doom of those highest ideals. How can ideals be reconstructed? Certainly not by some higher ideal, for it is precisely that highest ideal which we need most to have reconstructed. How then? The only way to reconstruct ideals is by certain data of experience. Experience alone can show us the inadequacy of our ideals and provide us with the materials and hints for the construction of more adequate ones. There is nothing more foolishly sentimental than persistently to cling to some "highest ideal" merely because it is so abstract and remote that it cannot be put to the test of experience in such a way as to make a fool see its insufficiency, although its derivatives and corollaries, the so-called lower ideals which can be tested, are constantly revealing their flaws. We claim that religious experience is of such a nature that it preëminently provides us with those data and that mental attitude by which we are able to fashion new ideals, and do this progressively to the end of continuous growth.

The popular present-day identification of religion with ideals makes God an ideal or system of ideals. There are some who frankly say this. Others are too confused in their thinking, or too much under the bondage of traditional concepts, to see that this is actually the attitude they have assumed even though they will not explicitly say so. E. S. Ames is a very good example of one who identifies religion with the pursuit of ideals and God as a symbol of such highest ideals. We believe he recognizes very clearly and definitely that this is his own position. Yet he constantly states his views in a manner that is confusing. This is very natural, perhaps necessary, for one who takes this view. For if God is a sort of glorified Santa Claus, serving to symbolize a spirit, a desired sys-

tem of habits and institutions, "social values" or
whatever other name one prefers, it is plain that
He will do this much more effectively if we truly
are rather confused in our thinking and half the
time take Him for a real person. It will help immensely if we can confuse the issue, befuddle ourselves and others, and so give the symbol the value
of a living person. Ames describes our idea of
God thus: "It is sometimes said that the God-
idea belongs peculiarly to the realm of values
rather than designating factual reality. But the
distinction between value-judgments and factual
judgments is not absolute. That it is a relative
distinction may be seen in the universal and inher-
ently teleological character of thought. All thinking is normally purposive." [1] This general principle we certainly would not dispute. But it is plain
that a very useful and beautiful confusion in our
ideas of God can be produced if we insist that God
as value-judgment is somehow also a sort of fact-
ual-judgment; and if we cherish Him as an ideal,
He is also a sort of fact. Of course the same is true
of Santa Claus. But Ames develops this cue immediately. He says:

> Only when extremely abstract and partial can
> it (thinking) be characterized as merely descrip-
> tive and factual. Again the God-idea is formed in
> terms of personality. And the conception of per-
> sonality involves primarily purposive action, not
> static being. The character of a person cannot be
> thought of except in terms of what he does. The
> idea of a supreme Person necessarily involves in the

---

[1] Ames, E. S., *Psychology of Religious Experience*, pp. 318, 319.

highest degree the element of will, of purpose and of movement toward great goals. It is a contradiction in terms to conceive a person as mere existence, that is, as fact simply. . . . The only kind of thinking of which human beings are capable is that which refers to ends, to needs, values. The God-idea is a teleological idea, and in being such it shares fundamentally in the nature of all ideas.[2]

Here is what seems to us a confusion of thought, no doubt brought on by the subconscious propensity to revere an idea as though it were a person when it is not. Of course an idea is teleological; but the teleology of the idea is not identical with the teleology of that to which the idea refers. The fact that my idea of a horse is tclelogical does not make the teleology of my idea identical with the teleology of the horse. If it did, beggars would ride. A horse, as a living organism, is of course teleological. So also is my idea of a horse teleological. Is then, my idea of a horse an adequate substitute for an actual living horse, because both horse and idea are teleological? Imagine one trying to comfort a poor lost wanderer in the wilds with the fact that his idea of a horse is as good as a horse because his idea is just as teleological as a horse could be. This is a juggling with words. Because a personality, and hence the supreme Person, must be purposive, and my idea of the supreme Person must also be purposive, therefore my idea of the supreme Person must be almost the same as an

_____
2 *Ibid.*

actual living supreme Person. For do not they both involve "primarily purposive action, not static being?"

We have found it necessary to dwell at some little length on this point because it is a form of confusion that seems to pervade so much of our modern popular views upon religion. We feel that such thinking would lead to the destruction of religion if religion depended on thinking—and it does in part depend upon thinking.

IDEALISM

There is, however, another and very different way in which God is identified with the Ideal and religion with the pursuit of ideals. It is the way of philosophic idealism. It holds that concepts, ideals, all universals, are present, objective beings, which are quite independent of our thinking; which may be discovered, cherished, adored, sought, by us, but are by no means merely creatures of our own minds. And in this sense, say these thinkers, God is an ideal or system of ideals.

This throws us back into the whole great problem of what is a concept and what is its status. Suffice it here to say that if concepts are held to make up a sort of immaterial spiritual world of themselves then that world must include every concept whatsoever, false as well as true, and every ideal, good and bad, along with every vile and loathsome thing that enters the human fancy and every horror that ever has or ever can come to our minds. No one could identify such a "world of pure essences" with God. It must be some selected portion of this "world" that is God. And then we are thrown back on the old problem of what universals, what ideals—among this infinite total-ity—constitute God? The ideals I esteem highest? —or those you exalt? The ideals that I thought best last year, or now?—or those that I shall hold highest ten years hence?

This identification of God with ideals has so many different forms and ramifications that we beat the air in our attack upon it unless we fix on some particular presentation of it. Let us take one of the most competent advocates of this religion of ideals and examine his case. There are a host from whom we might select but scarcely any more excellent than Pringle Pattison.

In his *Idea of God* (page 246) he says:

> Whence, then, are these ideals derived and what is the meaning of their presence in the human soul? Whence does Man possess this outlook upon a perfect Truth and Beauty and an infinite Good-ness, the world of empirical fact being, as Bacon says, in proportion inferior to the soul? Man did not weave them out of nothing any more than he brought himself into being. "It is He that hath made us, and not we ourselves"; and from the same fontal Reality must be derived those ideals which are the masterlight of all our seeing, the element, in particular, of our moral and religious life. *The presence of the Ideal is the reality of God within us!*[3]

---

3 Italics mine.

One strain of argument in this passage seems to be that because God made man and man has ideals, therefore ideals are of God. On the same line one could say that because man has errors, illusions, evil wishes and evil deeds, these also are of God. Or, pressing the matter further, because Man is not ideal and God made man, God must be other than ideal. "From the same fontal reality must be derived" everything in man. He speaks of man's outlook upon a perfect Truth and Beauty and an infinite Goodness. But just exactly what is meant by a perfect Truth, etc.? We admit that there is a common urge, an ultimate craving, common to all men, which is a sign of God working in us. But to call it an ideal is to falsely represent it. Our ideals are our attempts to interpret this urge, to conceive in thought its objective and to shape our conduct in such a way as to attain it. But these ideals are not the actual object which stirs us and can ultimately satisfy us. Truth for me is my knowledge of the facts. Perhaps by perfect truth the author means my knowledge of the total fact. But is God identical with myself in that status of knowing the total fact? When I worship God am I worshipping myself in that idealized state of knowing the total Fact? No, of course not, our author would reply. God does even now know the total fact, *i.e.* knows Himself completely. Why then speak of God as an ideal? Because I am religious only as I strive to be as God and like Him know the total fact. If that is what is meant then we must distinguish between two things. To confuse the two, as we believe our author does, is to render our thinking ambiguous. We must distinguish between (1) God the actual present total knower, and (2) ourselves in the prospective ideal state of knowing in like manner. The latter may be an ideal. But the former, God Himself, is not. What we worship, what we love and experience religiously, is God and not the ideal in any ordinary sense of the word ideal.

Consider another quotation:

> The human idea of God or of perfection is . . . . one which grows with man's own growth, acquiring fresh content from every advance in knowledge or in goodness, opening up fresh heights and depths to him who presses honestly forward; but he who penetrates farthest will be the last to say that he has attained. We are never at the goal, but as we move, the direction in which it lies becomes more and more definite. The movement and the direction imply the goal; they define it sufficiently for our human purposes; and in direct experience we are never at a loss to know what is higher and what is lower, what is better and what is worse.[4]

We are never at the goal, to be sure, but are we ever at God? Do we ever attain God? Yes, our author would reply. Then the goal plainly cannot be identical with God, the goal being ourselves in a certain anticipated situation or status, and not God. And when he says that in direct experience we are never at a loss to know what is higher and

_____
[4] *Ibid.*, pp. 248, 249.

what is lower, we cannot follow him.   To have direct experience of God does not mean to know what God knows.   It does not even mean that we ourselves have any adequate idea of God.   Direct experience is not knowledge.   It does not require great knowledge to experience God, and great knowledge does not result necessarily from the experience of God.   And there is no direct experience whatsoever, moral, religious, aesthetic or what not, which is of such a nature as to leave man "never at a loss to know what is higher and what is lower." By a number of carefully regulated direct experiences, one can test his ideas and attain a high degree of certainty.   But there is no unique sort of experience that in some mysterious way renders the mind infallible.

"Hence the ideal is precisely the most real thing in the world; and those ranges of our experience, such as religion, which are specifically concerned with the ideal, instead of being treated as a cloud-cuckoo-land of subjective fancy, may reasonably be accepted as the best interpreters we have of the true nature of reality."[5]

Here again we find the confusion between experience and knowledge and between the ideal and the living presence of God.   We experience reality and we interpret that experience.   The interpretation may be false or true, or partly both.   But an immediate experience cannot be either false or true. It is simply so much given data of experience. Some of our ideals may be the best interpreters we

---

[5] *Ibid.*, p. 252.

have of our immediate experience of reality; but to say that ideals, merely because they are ideals, regardless of their nature, are the best interpreters of reality, is plainly false. Some ideals are good interpretations and some are not; none are infallible, and many flatly contradict one another.

We know that Mr. Pringle Pattison would say we are not fair to him; that he does not mean what we say he means. We would agree with him. He does not mean what his assertions imply. That is precisely our criticism of his thought. We feel there is a confusion in which his meanings become entangled. This confusion is so wide-spread in modern religious thought, is doing so much damage to the interests of religion, and will do much more if continued, that we feel it necessary to criticize it very severely. We believe we have very much in common with Mr. Pringle Pattison. It is his method of treating the subject matter rather than his apprehension of the subject matter with which we disagree, if we correctly detect his apprehension under the treatment.

We hold that an ideal is nothing else than an hypothesis until it becomes a demonstrated fact; and then it is no longer an ideal but an existent situation or status. It is true that one of the chief functions of religious experience is to quicken new aspirations, new visions of some desired situation, social reorganization, habit system, world outlook, or what not. But these visions and aspirations are hypotheses, programs of action, no more divine than any other, and should be subjected to all the tests, criticisms, analyses and tentative experiments that any other hypothesis is subjected to. They are not divine, but our reactions to the divine. It is quite natural that when religious experience leads to the condemnation of established ways and stimulates the constructive powers of the imagination, there should spring forth, as though from some mysterious outside source, marvelously accurate insights into the better way of life. But if we worship this insight or vision as though it were God, we are idolators. Much of the worst kind of dogmatism has come from this deification of doctrine and vision. Our thoughts and our ideals are never identical with God.

RELIGION AND GROWTH

It is this identification of God with certain ideals, scientific and philosophical systems, institutions and practices, which has frustrated religious experience in the performance of its genuine progressive function of leading human life to ever greater abundance. It is this that has made religion so often the chief opponent to new scientific theories, new and better fashioned institutions, new moral interpretations, and new and better ideals. It is this theory of religion that has perverted the whole function of religious experience. The great religious leaders of history have been able to break through the obstructions of established ideals and institutions, declaring: Ye have heard of old time, but I say unto you. It was thus all the great Hebrew prophets spoke. It is thus that all the outstanding religious leaders have spoken. And in more humble fashion ordinary religious men have come from the hour of religious experience with a new and different vision.   This is a part of what is involved in conversion.

The age which thinks that the chief work of religion is to whip or lure men into fuller devotion to established ideals has forgotten the greatest of all religious injunctions: "Ye must be born again." It is because religion is independent of any and all conceived ideals, that it provides the way by which men can transcend their limitations, and rise beyond any and all conceived ideals to others that are more adequate to the demands of a changing human life.   All our ideals, all our cherished values, are constructed out of our judgments, which always contain some error when applied to the vast, complex, concrete affairs of daily life.   They not only fall short of the best; they have in them that error, that divergence from rectitude, which is the seed of death.   That slight divergence, if too long continued, leads to death.   They may be the best we have, excepting only God; but they are not God. Religious experience provides a way of salvation because it is a way up and out and beyond our ideals.

OBJECTIVES VERSUS IDEALS

For the most part tradition and illusion have shaped our living. We prate much of "ideals" but they have played only a small part in determining the manner of our response to the stimuli that

assail us and that organic process of adapting ourselves to environment which we call living.   The
most popular "ideals" about which we hear most
talk, belong with the illusions.   As illusions they
do shape our lives, but this is not what is ordinarily
meant by the shaping of life through "ideals."   As
illusions they enable us to conceal from our own
consciousness the sordid motives which may actu-
ate us.   Or they serve to screen our consciousness
from disagreeable facts in the environment.   They
help us to bolster up our self-respct;   to glorify
ourselves and others;   to see things in a rosy light.
It is plain, then, that these illusions, some of which
are called ideals, and others not, do play a very
important part in our living.   Tradition, working
in us in the form of habit, drives us on;   but illusions brighten the path before us and make the
going far more pleasant.   Indeed, the false rosy
light of illusion may make the path of traditional
habit quite exhilarating.   From earliest infancy the
traditions of our time and people mould the organic
process of our living into the form of certain habits.
So our way of life is determined.   But above us
float illusions, some of which we call ideals, others
we recognize to be nothing but pleasant fancies,
while others we mistake to be facts.   They are as
a pillar of cloud by day, and of fire by night, which
float before us.   Our system of habits is undoubtedly modified to some degree by these illusions,
sometimes for the worse, sometimes for the better,
But it is perhaps fortunate for us that the organic

process is not too greatly under the control of these illusions.

What is the difference between an objective and an ideal? When an ideal begins to determine the actual organic process of living, so that one breathes, eats, digests his food, moves, sits, and stands in such a way as to sustain it, it is an objective. But a great many "ideals," perhaps most ideals, as they are popularly conceived and portrayed, never could be objectives because, if the organic process were shaped by them, life would end. The vital status could not be maintained. Or, if the animal objectives could still be maintained, the human objective of enlarging the fullness and range of environment to which adaptation is made, would be rendered impossible. In short, a great many ideals as popularly conceived, are impossible, not because they are so "high," but because they are so worthless. They are pleasant dreams with which to beguile the tedium. They are pleasant not bcause thy are so "lofty," but because they provide a fairyland where certain of our repressed impulses can find imaginary fulfillment. These "ideals" are the stock materials with which some orators, some writers and artists entertain the crowd. They enable us to forget our unpleasant surroundings, much as a drug might do. They "inspire us," they exhilarate us, they console us, but they do not continuously and fully determine our striving and our doing. They are the ice cream and cake of life. But the bread and meat consist of our objectives, both the animal and the distinctively human.

But when a civilization reaches maturity, as ours has done, and as the Greco-Roman did, it calls for something more than ideals that entertain and inspire. It calls for a clarification of those facts by which and for which we live. It calls for objectives. As long as tradition sustains us, we can get along without knowledge of objectives. We can give our thoughts to the luxury of ideals; and because of the meager, mean and impoverished life which tradition often imposes, these pleasant, fanciful illusions are indispensable to make our living humanly tolerable. We must deceive ourselves with "ideals" in order to carry on. But when tradition breaks down, and we must direct our own life according to goods which we have deliberately chosen —when we thus pass from drift to mastery—we must know our objectives. We must know that actual good, the conservation and increase of which constitutes the good of living. Pleasant "ideals," illusions, and dreams, however valuable in the past, no longer suffice.

WHAT IS AN OBJECTIVE?

By objective we mean, not an end result, but some satisfaction which is maintained, and may be increased, by those activities which we call living. For example, a certain quantitative proportion of oxygen and carbon dioxide must be kept in the air cells of the lungs. Our behavior, our rising, walking, sleeping, eating, heart beating, blood circulation, etc., are adjusted in such a way as to maintain this composition of the air in the air cells. This is plainly an objective of living in the sense that all our activities are controlled in such a way as to provide this condition. Reacting to circumstances in such a way as to maintain this condition is what is called adaptation to environment. Also a certain temperature of the body must be maintained, and all our going and coming is adjusted in such a way as to keep this temperature without variation. This also is an objective. The same is true of a certain chemical composition of the blood. Here again all our organic processes, the metabolism of our food, as well as our thinking, our loving, our dreaming, our playing, our fighting, etc., are as a rule, adjusted in such a way as to protect and sustain this condition. The same is true of the integrity of our organism. We strive to keep a whole skin. Our activities are all adjusted in such a way as to preserve the organism from harm. So this also is one of the objectives of living. So likewise with other vital conditions. The objective of living is to provide, preserve, and magnify the conditions which living requires. The objective of living is to preserve and magnify itself.

What we have described are objectives of animal living, and humans share these objectives with the lower animals. But humans have also other objectives. We call the man-beast human because of that way of life which he sometimes pursues, and to which he may give himself in ever larger measure under proper nurture and favorable conditions. Man is sporadically human, but not consistently so. The objectives, then, which characterize human living, must not be represented as controlling the whole life of man; and at times they do not control any part of his living. At times, and perhaps for more of the time than is commonly thought, the purely animal objectives are sufficient to account for all that he does.

The animal objective is to preserve life while the human is to develop more abundant life. To live means to react to environment in such a way as to maintain such vital conditions as air in the air cells, a certain composition of the blood, temperature, organic integrity, etc. To live more abundantly means to maintain these vital conditions in reaction to a more ample environment. To live humanly is to engage in the progressive organization of an ever more complex system of habits by which one reacts to an ever larger portion of the world. Distinctive human behavior is that continuous variation of behavior which provides for the needs of the organism in adaptation to an ever wider and fuller environment. The animal reacts to the environment only in so far as necessary to meet the needs of the organism; the human meets the needs of the organism in order to react to ever more of environment. This difference holds true, of course, only when the human is truly different from the lower animals. But he is not always different.

OBJECTIVES VERSUS TRADITION

During most of the time our race has lived on this planet men have not asked why they followed the beaten path before them, nor whether there were other better ways, nor whether the labor of going on was worth the cost, nor what the goods they sought. It was the path of tradition they followed; and the system of habits, ground into them from earliest infancy, kept their feet in the narrow way. "Theirs not to reason why, theirs but to do and die." The tradition they followed was gradually accumulated throughout many generations without any particular plan or purpose. Or perhaps it would be more correct to say that the tradition had been developed throughout many years under many different conflicting plans and purposes. It always served in a general way to preserve the biological existence of those who followed it; for those who followed traditions that did not do this perished, and their traditions with them. No doubt these traditions provided for many other goods over and above the biological, but this they have never done consistently nor in the most economical fashion. Yet most men in all times and among all peoples have followed some such tradition.

On either side of the beaten way of tradition have been many pleasant things to be had. But for the most part men have turned neither to the right hand nor to the left. They have not even seen these pleasant things. They have seen only the road straight in front of them and the dust beneath their feet. They have been too busy to note anything else. They have been carrying such heavy burdens, or have been so sorely pressed to keep from being trampled under foot by the throng, that they have not had opportunity to lift their eyes and look around. They have not asked: Wherefore all this suffering and doing? What good are we conserving or creating? They have simply stolidly gone their way. If ever the question, What's the use? rose to the fringe of consciousness, they put it quickly from them. Such a question is too disturbing. It destroys all contentment in living. It takes the zest and force from all doing. It sometimes leads to suicide and insanity. It makes for social unrest and impairs the smooth working of the social machinery. Hence the question is abhorred both by those who profit by the established social order and by those who are oppressed by it.

But there have been a few rare times when the burdens borne have been lightened by machinery or slavery; and the absorbing attention to keep from being trampled under foot has been relieved by reason of a more stable social order and protection from foreign aggression. If such a time of prosperity and peace happened to coincide with a certain ripeness of civilization, in which the accumulated experience of the race could be transmitted to the individual in such manner as to give him insight and reach of thought, the great question has been raised: What is the good of living?

Why this sacrifice?   Why this devotion?   Why follow this inconvenient or devious route?   Why this custom or this form of conduct?   Why suffer and toil?   So arises the age of sophistication.

Now as long as this question concerning the values and objectives of human living is limited to a few only, who are very old and experienced, or have a peculiar philosophical bent, or are especially provided with leisure, it is not a serious matter.   As long as tradition serves to shape the lives of the great mass of men during youth and middle man-hood, the question concerning the objectives of human living is largely an academic one.   But when sophistication reaches any large number of men, especially the youth and middle aged, upon whom rest the loads of life and the control of the social process, and when these begin to criticize the traditional way, asking the why and wherefore of it, and examining to see if there be not some other better way, then the question concerning the good of living becomes a very urgent one.   Then tradi-tion begins to crumble and fails to serve as the stay of life.

Our Western civilization, perhaps one may say the world's civilization, has reached such a time as this or is fast approaching it.   Traditions still hold, of course; but the flood that threatens their stability is trickling over the top of the dykes, and beneath it the dykes are crumbling.   The opening may widen and deepen and the torrent pour through. After us, perhaps, the deluge.   The halo that once hovered over the state, the home, industry, and the country, is gone.   It is shown to be an illusion. The great values which once we thought we saw in these things are torn away in modern essay and fiction.   There is for the present a zest and joy in tearing away fictions and laying bare the facts, unadorned and indubitable.   But will this zest and joy continue?   Not if the zest and joy consists merely in tearing away of the illusions, for in the course of time there will be no more illusions to tear away.   The joy of living can continue only if in the facts themselves can be found absorbing values.   But if the childish delight of destroying shams is our only delight, our end is near.

Now in such a time as this, ideals cannot be worshipped simply because we must commit our lives to them and thus bring to light the inadequacy and illusion that is in them.   As long as we committed our lives to tradition we could worship ideals and identify religion with idealism because the ideals were not put to the test of actual life.   In our confused thinking the ideals could serve as symbols of God;   and as the idol worshipper may truly worship God through the symbolism of his idol, not thinking clearly enough to make distinctions, so we also could worship God through our ideals.   But there comes a time when idol worshipping is dangerous and destructive to genuine religion, just because people begin to see through it;   and there comes a time when ideal worshipping is dangerous and destructive in like manner.   We have reached such a time.   The worship of ideals is a form of idolatry that can no longer satisfy the need of our time.   The ideals that inspired a decade ago have turned to chaff in our mouths.   We have a religious hunger, and if it be not satisfied we are undone.   We demand a degree of religious certainty such as no other time ever demanded.   With the new scientific method that has come into our possession and our present ripeness of experience and sophistication, we are overhauling all our old beliefs and holding up to scorn all the old idols and bugaboos that charmed or frightened into conformity a more credulous age.   Nothing but religious certainty will satisfy us.   Where and how shall we find it in the light of modern psychology and scientific method?   That is the problem we have been considering in preceding chapters.

CHAPTER XI

CHRISTIANITY AND PSYCHOLOGY

Perhaps the two most striking influences at work today in the field of the psychology of religion are the Freudian theories and the work of William James. We do not mean that they are the most valuable. Certainly the Freudian theories, however illuminating they have been in bringing human nature into the light, have been full of quite fantastic ideas, undisciplined by scientific method. But because of the great influence of these two strains of thought we must consider them at some length. While our criticism of the Newer Psychology, as it is presented by some of its advocates, will be quite severe, we should be greatly misunderstood if it was thought we did not appreciate the great service these theories have rendered in giving us a better understanding of the human being. Indeed we think they have rendered an inestimable service.

Vagrant impulses, old habits, and instincts do certainly at times manifest themselves in our dreams, and at other times when we are off our guard. Because these characteristics do display themselves at such times, it has been maintained that these impulses are always in us; that they remain suppressed in some unconscious region of our natures, (in the "unconscious mind") and spring forth in sleep or in the moment of shock, because the system of habits which ordinarily suppresses them is then relaxed or disrupted. But this occasional breaking forth of impulses in dreams and shock does not necessarily prove that these impulses continue all the time in some repressed region of mind, organically active but prevented by the censor from shaping behavior and consciousness. On the contrary it would seem to prove that, when we are asleep or at other times when these errant impulses show themselves, the organic system of response that controls our normal life, has become temporarily disorganized. Excitement, weariness, sleep or illness or shock may disrupt the controlling system of response which we commonly call the prevailing character of the man. With this disruption, certain responses are released from control and operate independently. Being no longer under the control of the organic system of conduct, these released responses may follow old "paths." They may revert to the form in which they were originally established. If as a child I learned to hang my hat in a certain place, I may under stress of great excitement, or somnambulism, try to hang my hat at the old place even though I have been hanging it elsewhere for many years. This does not mean that all these years I have suppressed an impulse to hang my hat there. It only means that the present crisis has disrupted my system of habits and this particular response, when thus thrown out of the system, reverts to the old course of procedure. Hence at these times, when the system of habits are disrupted, we have reversions to the primitive, the infantile, or to the behavior of any bygone time that has made a deep impression upon us. Of course this disruption may not be limited to special occasions but may be more or less chronic. Chronic disruption, however, is not so widespread among normal people, we believe, as some Freudians seem to teach.

The whole matter may be summed up by saying: Habitual responses fall into a system, but in so far as it is a distinctively human system, it is constantly developing into greater complexity. This means that new impulses are aroused, and old ones modified, by reason of our response to additional features in the total situation to which we are adapting ourselves. These new impulses must be assimilated into the system of habits. If they are not assimilated, or until they are assimilated, they conflict with the established habits. Now this conflict may be overcome or avoided by the continuous adjustment of these new impulses to fit the system of habits. This is what is meant by assimilation. This means the progressive organization of an ever more complex system of habits. This means coming into converse with ever more of the fullness of the world. This means progressive adaptation to an ever more ample environment, entering into a fuller life.

But this continuous readjustment and assimilation of new impulses into old systems of habit rarely is perfect in its operation. Some new impulses resist adjustment to the old habits; and some old habits are too rigid to be modified in the manner required for assimilation. Then occurs conflict. Sometimes the impulse is suppressed. Sometimes certain habits are suppressed. Then arises mental aberation and distress of mind. There is loss of poise and peace. One finds himself working against a sort of mental friction. Efficiency is impaired. There is a strain and tension about the organism which bodes ill for happiness and power.

All the responses or habits that go to make up the life of an individual may be divided into the coöperative and the antagonistic. Coöperative are those that work together in a single system. The antagonistic are those that conflict with this system. The antagonistic habits may themselves form a system which runs counter to the first, or they may be merely vagrant impulses. In extreme cases the counter system constitutes what is called a dual personality.

Now these conflicting impulses or habits, if persistently prevented from fulfillment because of their antagonism with the prevailing bent of the individual, may distort one's thinking and perception. Above all they may shape one's dreams. They may cause one to build air castles, which must be very carefully distinguished from that constructive work of the imagination in which one maps out programs of action and interprets the immediate data of experience. These air castles are not programs of action but are the exact opposite.

They are constructed precisely in order to make programs of action unnecessary; to soothe and pacify these suppressed impulses which might otherwise disrupt the prevailing bent of character. Such air castles differ from constructive imagination in two ways. They are pacifiers rather than guides to action; and they serve to conceal the facts from consciousness rather than inform the consciousness concerning facts by suggesting verifiable theories.

Now of late it has become the vogue among followers of Freud and Jung and Brill to represent all religion, or the greater part of it, as such a work of phantasy. Religion, they say, is the great air castle, collectively constructed by the fancy of many generations, handed down for centuries, constantly readapted and embroidered by the fancy of each age, sustained by an army of officials and mighty institutions and always shaped to provide a fictitious fulfillment to impulses which are frustrated by the facts of life. It is the great game of bunkum which the human race has developed for the purpose of "kidding itself" into thinking the world is nearer to the heart's desire than in fact it is. It is the Great Myth which is dearer to the human heart than any fact could ever be.

Instead of treating this "psychology of religion" in a general way, we think it will add to clearness and explicitness if we give our attention to one typical exponent of this view. Everett Dean Martin has very recently expounded and defended this theory in *The Mystery of Religion*. Let us briefly examine this book. But let us say at the start, that while we do not agree with its major conclusions, there are certainly flashes of true insight here and there, and all that he says might well be applied to certain instances of religion (or pseudo-religion, some would say). For there is a romantic and sentimental religion just as there is a romantic and sentimental art, and love and history and patriotism, and sociology and psychology and everything else that man may undertake. For man does not leave his sentimentality and romanticism behind him even when he takes up the study of Freudian Psychology.

> The tendency has been too strong, especially on the part of thinkers who have grown up in Protestant communities and have been influenced by nineteenth century natural science, to regard religion as an affair of ideas. Traditional beliefs, dogma, and myth, which have come down to us from an earlier time, and are of course out of harmony, when taken literally, with the naturalism of our age, are regarded by these thinkers as merely crude attempts at a rational explanation of nature and life, made by simple people who, alas, could not know as much about the world as "scientific" moderns know. [But, says Martin, these fabrications are] cherished generation after generation to be an escape from the very reality into which modernism would fain now confine the objects of religious interest![1]

What Martin says is no doubt true of many

---

1 Everett Dean Martin, *The Mystery of Religion*, p. 9.

religious people.   But is it true of religion *per se?*
In answer to that question we must make certain
distinctions which Martin has failed to make.

It is true that these "beliefs, dogma, myth,"
cherished by religion and handed down by tradition are not bona fide attempts to "give a rational
explanation of nature" in terms of physics or chemistry.   They are not attempts to explain the data
of experience treated by physics and chemistry any
more than biology is such an attempt.   Biology
deals with different data from physics and chemistry and the "explanations" of the latter are not
applicable to the biological data.   Nor is the biological explanation applicable to the psychological,
because the latter seeks explanation of a still different datum.   Of course the different sciences can be
of enormous assistance to one another; but it is a
narrow and obtuse mind that tries to reduce one to
the other.   Now the "beliefs, dogma, and myth"
cherished by religious tradition are not attempts to
explain the data of any of these sciences.   They are
attempts to explain a wholly different kind of
datum, the datum of religious experience.   The
several recognized sciences can help immensely in
our attempts to understand this datum of religious
experience, but until we get a science which can
study this datum without reducing it to the data
peculiar to some other science, we cannot give a
scientific explanation of it, any more than physics
can give a satisfactory explanation of the living
organism, rendering biology superfluous.

It is for this reason, and only for this reason that

the effort "to square these old beliefs with modern knowledge" results in failure. It is not because these beliefs serve to delude, and men want to be deluded; it is not because religion is a fairy tale which men refuse to relinquish because it pleases them more than fact. But it is because these beliefs are sincere attempts, guesses if you will, to discover the true significance of certain data of experience; and they cannot be reduced to any of the recognized sciences of the day for the same reason that psychology cannot be reduced to chemistry pure and simple. The science adapted to the investigation of the religious problem has not yet attained maturity. It is still in the womb of philosophy, which is the mother of all the sciences.

Martin is perfectly correct when he says that religious beliefs cannot be reduced to sociological theories, as some would try to do. Sociology perhaps comes closer to the religious datum of experience than any other science, because its datum is the most complex. In the sociological datum is merged the data of all the earlier sciences. But the religious datum is even more complex than the sociological. and the reason why Christianity cannot be reduced to sociology is not because it is a cherished myth, as Martin asserts, but for the reason mentioned. In the heyday of physics there were certain moderns who tried to reduce Christianity to physics and taught that God was the Great First Cause. Of late, when biology and evolution have captured the imagination of the scientific mind, there are some who want to interpret God as the vital urge, the creative impulse in evolution, a biological entity. In other words, they try to deal with the datum of religious experience as though it were a biological datum. And now in our own time there are "hot little men" who would make the object of religious experience identical with society. But surely we will learn some day that no science can adequately treat data that lie outside its field. It is always a mistake to consider the raw experience entering awareness as "nothing but" the data of some special science. Raw experience, prior to analysis, is not a datum which any science can treat. Hence religious belief cannot be reduced to the terms of modern science. Martin is right in this, but his own theory is quite as bad.

We are sure that Martin is correct in the following statement:[2] "Whenever people discuss the question of religion as if it were primarily the problem of the nature and process of Creation—for instance, whether Genesis or the *Origin of Species* gives us a correct account of the appearance of the varied forms of plant and animal life, or whether the protozoan process of parthenogenesis may explain also the story of the virgin birth of Christ—you may be sure that they have forgotten for the time what religion is really about." With all this we heartily agree. But if we added the very next sentence that follows, we would have to disagree. Religion is not "about" the matters treated by modern science. But neither is it about a myth. Neither is it about something outside sense experience.

[2] *op. cit.* p. 21.

But the datum of sense which religion is "about" is too complex for any of our sciences to interpret. Hence the best we can do is to interpret our experience by some "belief." We may invent some new-fangled little belief of our own or accept a belief handed down from the greatest religious teachers of history, but some belief we must have if we are to get any meaning at all out of our experience, since no science is available. Martin makes a very interesting and suggestive comparison between the scientific concept, which he considers a symbol, and the religious belief, dogma, or formula, which is also a symbol. The analogy he uses is that of the map and the flag as being, respectively, symbols of United States.[3] "Each of these may be used to represent one country, to bring its reality to our minds. Each is a humanity-created device, the use of which makes it easy for us to conceive in some way of the reality for which it stands, and to behave in a certain way toward that reality. Yet neither is that reality, nor in fact at all like it." The map is a utilitarian symbol, enabling us

> to orient ourselves practically to the reality which we call United States. Yet the map does not represent the country any more truly than does the flag. The latter symbol—and none will deny that it is a symbol—stands for certain emotions and ideals. It is in these respects—and very real they are, too—that the flag means America to us. It is this America which inspires men's devotion and their pride in the country. . . . The flag

3 *op. cit.* p. 34 ff.

orients us emotionally to America. Both these symbols stand for the same America, yet they represent very differnet interests, and have different uses. The map might be termed the "scientific" symbol, and the flag the "religious" symbol. The map is an instrument of those adaptations to reality which consist of external movements, going to certain places, shipping materials, sending messages, etc. The flag stands for those forms of adaptation that emanate from within. Here, too, there is overt behavior, but it is, or should be, behavior dominated by unseen values, rather than by objective considerations.

He adds that the map involves directed thinking while the flag stands for free association. Directed thinking solves problems, discovers the correct means to desired ends, deals with a "world of objective realities." Free association is dreaming, the play of fancy, arising out of the conflict of wishes and the attempt to solve these conflicts by means of air castles.

We must study this interesting analogy with some care, for we believe it contains the thought of the entire book put into a nut shell.

In the first place the map does not symbolize the whole of United States. United States is not merely "a geographical expression." But the map does not even symbolize all the geographical features. A few—a very, very few compared to the concrete totality of American geography—have been selected because of their practical importance in guiding our travels; and these few features, these data, are symbolized by the map to the exclusion of everything else. In this respect the map is a perfect example of the scientific concept which, as we have seen, abstracts a few data from the wealth of concrete experience, and carefully plots the space-time relations of these to the exclusion of everything else.

But how about the flag? Does it symbolize nothing but illusions, fancies, and fictions? To be sure it does symbolize these. But is that all? Is the United States, symbolized by the flag, not a fact? Is it only a dream? Is it the product of "free association," a castle in Spain? Manifestly that is absurd. Part, at least, of that symbolized by the flag is out there in space and time, as truly as that symbolized by the map. But the symbolism of the flag differs from that of the map in two ways, which we must define and distinguish with care.

Whatever else the symbolism of the flag may include, it refers to an object, which is just as independent of the individual's dreams and wishes as those meager geographical features symbolized by the map. But the difference is that the fact or object symbolized by the flag is far more complex, immeasurably richer in detail and concrete fullness, than is the object symbolized by the map. The flag properly symbolizes the concrete totality of United States precisely as it is and without those exclusions and abstractions involved in the symbolism of the map. In this respect the flag symbolizes the "real" United States while the map does not.

But there is another respect in which it can be said that the flag does represent myth, fancy, and guess, while the map represents things as they are. The concrete totality of United States is far too complex and vast for us to cognize. We have no adequate concept with which clearly to think it. We cannot doubt that it is; but just what it is we do not know. Most of our concepts by which we think it must be guesses, theories, and dreams, because we have not yet developed the scientific method which is adequate to test the correctness of our ideas of such a complex totality. Consequently our ideas about United States, both its present status and its future destiny, must contain a great deal of error. And where scientific method is not available to correct our thinking, there will certainly creep into it a great deal of distortion due to our own wishes.

But right here another important distinction must be made. It depends on the individual whether his ideas concerning the United States, symbolized by the flag, are tentative but sincere theories concerning the character of his country held subject to verification, or whether they are illusions carefully preserved in the face of contrary evidence. I think we are all well enough acquainted with these two kinds of patriots. None of us are altogether free of illusions to which we cling, especially where scientific methods of verification are not available, or, for that matter, where they are; but to say that the flag is for all people preëminently a means of sustaining certain illusions and concealing fact, is certainly a gross error.

Now all this applies precisely to the difference between the scientific concept and the religious belief. It depends upon the individual whether his religious symbols are cherished for the sake of blinding his eyes to the truth; or whether they are his sincere and earnest gropings after the true significance of certain data of experience which are so complex that no scientific method is yet able to distinguish with certainty between truth and error. But the object of religious experience is just as indubitably an object, independent of human fancies, as is the United States. Just what it is, we cannot be sure; but that it is, we cannot doubt. The religious datum is given; and it is more certain than the seemingly verified concepts of science. In this sense the religious symbol most certainly does not represent an illusion or a dream.

We need not follow Mr. Martin through all his expositions of religious doctrines, ceremonies, hymns, prayers, institutions. He applies all the elaborate paraphernalia of psychoanalysis and Freudian psychology, the Oedipus complex and Electra complex, the libido and its regression, infantilism, the Father image and Mother image, etc., etc. The church is symbolic of mother and symbolically satisfies the craving to return to be with the mother, if not actually in the mother, which is said to be an almost universal trait of human beings. The erotic motive is found in such songs as, "Jesus, lover of my soul, let me to

Thy bosom fly," "I heard the voice of Jesus say, come unto Me and rest, Lay down, thou weary one, lay down thy head upon My breast."

> Leave your sins and come to Jesus;
> He will enfold you in his arms.
> In the arms of my dear Saviour
> O there are ten thousand charms.

In all this he sees a symbolic imaginative fulfillment of the sexual impulse. Indeed he thinks it is so indubitably manifest that it is beyond discussion. No doubt his interpretation is correct with respect to individual cases. But here again we see too wide generalization. Unquestionably these words are symbolic. But do they always symbolize the same thing? As a fact unscientific symbolism, (and this is certainly unscientific), is the most variable and vague pointer imaginable. Different people mean most different things when using the same words. But such symbolism is far more variable between individuals, and in the usage of the same individual at different times, than are ordinary words.

Perhaps the background of Mr. Martin's thought is best revealed at the beginning of the last chapter. After describing the aspect of land and sea on a beautiful August afternoon he says:

> Why cannot the life of man be like this? Why must we invent fictions in order to live, in order to find the meaning and value of our world? Why are we different from those flowers which fill this summer day with their own fragrance and

color, or those waves which fall upon the shore, breaking and receding and content with their inevitability? . . . Other things in nature are complete; they are what they are; we are not. We must find the meaning and value of our lives in fiction and illusion. We must find escape and compensation where other living things are content with reality. Reality for us is but half hospitable. Over it we have woven the web of civilization and set the ends of self-consciousness, and between that which man has achieved—and necessarily so—and that which he is by nature, compromise must be made. There is no return to nature for us. The meaning of life for us is no longer to be realized in the mere fact of living. We must create.

Here he has raised most interesting and profound questions. His failure to answer them, is what has led him so far astray in his attempts to interpret religion. Or rather his failure to discover the true meaning of religion is what renders these questions so baffling to him. If he knew the nature of religion, he could solve these problems. Let us consider them, for we believe they lie at the root of his whole difficulty. Since his psychology makes human nature appear more incomprehensible than ever, there is plainly something wrong with his psychology.

Why can man not be content to deal with things as they are, as do the flowers and the beasts? We might answer the question in many different ways. But let us keep as close to biology as we can. It is because man is sensitive to so many different stimuli from the environment that he is confused. Or, putting it in other words, it is because so many different impulses are awakened within him under normal conditions that they conflict with one another. Or, putting the same matter in still other words, it is because he is physiologically so constituted that he must constantly react to an ever wider and fuller environment. It is because of this that his deepest urge is for adaptation to a more diversified and ample world. His most characteristic propensity is toward more abundant life.

It is a mistake to think that man primarily flees from reality, that he first of all craves to ornament the world of fact with fictions. The fictions of the child, for instance, are not cherished illusions. They are very different from those hallucinations which the psychiatrist finds in the diseased mind. The fictions of the child are his groping efforts to deal with that large world of fact which he discerns round about him but which is beyond his reach. His fictions, instead of being flights from reality, are his nearest approaches to reality. By means of them he is trying to get closer to the world of fact, rather than farther from it. No, man's bent is to bring his multiplying impulses to fulfillment in reaction to the multiplying stimuli that assail him. But this is an exceedingly difficult undertaking. If he is successful he develops an ever more complex system of habits in adaptation to an ever widening and diversifying environment. But if he is unsuccessful, either one of two things may happen. But before mentioning these two possibilities, we must understand that success is always a relative matter. No man is absolutely, unqualifiedly successful. And perhaps all men share to some degree these two forms of failure.

One form of failure is to have the multiplying impulses deadened; to allow the imprisoning crust of routine habit gradually to harden; to degenrate into an automatic system of habits, adapted, it may be, to a few elements of the immediate environment, but unresponsive to anything else. One who suffers this failure, becoming a brother to the ox, does truly approximate that blissful state of the flowers and beasts. For their contentment and peace amount to nothing more than this; that their unchanging responses are adapted to a few meager elements of their immediate environment to the exclusion of everything else. They are alive only to this minimum degree, for to be alive means to respond; and they respond to the very fewest possible stimuli necessary to escape destruction and decay. They are content simply because they have no urge toward more abundant life.

The second kind of failure is the fiction-making of which Martin has so much to say. It arises when man is unable to bring all his impulses to fulfillment, precisely because they are so many and diverse. He must content himself with fulfillment of those which he has been able to harmonize, suppressing the others. These suppressed impulses do give rise to constructions of the fancy, to illusions and distortions of thought.

But it is when we come to the relatively normal and successful man, that the Freudians are most at fault. This mistake is to think that a fiction is always constructed and cherished for the sake of the fiction. On the contrary a fiction, first of all, is a theory. It is a groping out into the unknown. It is an attempt to discover that bigger fact which encompasses one, and which one has experienced, but has not clearly cognized. There is no other way of discovering that bigger fact, there is no other way of adapting oneself to that larger, fuller world that is about, except by means of such theories. One must first form an hypothesis subject to verification. But with respect to the most of our theories we do not yet have a scientific technique which is able to test their truth. That being the case there are just two alternatives open to us. One is to cease to hold any theory at all about such matters, and to exclude them altogether from consideration; to settle down to a humdrum existence in adaptation to those few elements of environment which are most immediate, simple and certain. In other words, cease to strive toward the more abundant life except where assured scientific method can guide us. The other alternative is to strive constantly toward those inadequately known but experienced facts; to hold to whatever theory will keep one in converse with them, until one can get more light.

Now of course there are all degrees and gradations between such theories, which are sincere gropings after fact, and mere phantasy which is cherished for its own sake and for the purpose of concealing facts. No doubt there are many cases where men cherish the fiction for its own sake, because it is a fiction, and because they do not want to know the facts. But this is not nearly so common as some would think. And often the surface appearance will give that impression when a deeper study will show that the individual is truly reacting to certain facts by means of a very inadequate theory; and his tenacity to this theory is not due to love of fiction but to unwillingness to allow this dim fact to slip from him forever, this theory being the only hold he has upon it. We who may not have any sense of that fact, or, more commonly still, cannot see how that theory serves to reveal that fact to him, think that he is stubbornly holding to a fiction because it is a fiction and nothing else.

Now this significance of theory, especially when applied to the object of religious experience, Mr. Martin and the Freudians seem to have overlooked quite completely. There is some excuse for this misunderstanding of religious people because these people are often unwilling to admit that their beliefs are theories. They insist with extreme dogmatism that their beliefs are absolute certainties. But there is good reason for this. The Object which these people experience, and which they strive to cognize through such beliefs, is not accessible to them by any other beliefs. It becomes, then, for them, a matter of giving up all converse with this object or else holding on to the belief, however crude it may be. Until we have some method for adequately testing by scientific methods our beliefs concerning the object of religious experience, this attitude is likely to be common.

But most important of all is the fact that these dogmatic religious people often use these beliefs as symbols or guides in entering into the mystic experience of God. The beliefs then become instruments by which they enter into the experience of a certain object, instead of theories for cognizing that object. Their training and habits may be such that they cannot enter into this experience in any other way. These symbols are devices by which they throw themselves into that mental attitude of wide-open awareness which we have seen to be requisite for the experiencing of God. The psychology of such mental devices is perfectly intelligible and in no way militates against the "reality" of the ensuing experience. Hence the tenacity with which such people hold to their beliefs, dogmas, and symbols of whatever sort, is not a case of mulish stubbornness, much less a love of fiction for its own sake; but it is a clinging to an indispensable instrument. They hold to such beliefs as a householder might hold on to the only key he has by which to enter his home. It is not a matter of the key's being false or true. False and true do not apply at all, even though the person involved may think it a matter of true and false. It is simply a matter of keeping the only key he has for opening the door into a certain region of experience. This experience is experience of fact; but it is not knowledge of fact.   There is no way to certain knowledge save by verified theory.

We feel that Mr. Martin misconstrues the nature of religion.   But in doing so he has merely followed a prevalent Freudian view of religion.

WILLIAM JAMES

Let us now turn to the psychology of Christianity represented by William James.   Ever since his Gifford lectures on "Varieties of Religious Experience" he has probably shaped the psychology of religion more than any other.

The wonderful thing about William James, that makes his work of such incomparable value, was that intellectual honesty that enabled him to observe accurately and record fairly all the facts that came to his attention, without allowing his own theories to distort or obscure them.   There was about him an intellectual humility that was also the highest kind of nobility, for he would present those aspects of the matter which revealed the inadequacy of all his own thinking and then would calmly admit that there were mysteries he had not begun to fathom.   And he would throw out suggestions that would have brought ridicule upon a smaller man from the representatives of scientific thought;   not because his suggestions were necessarily ridiculous, but because the advocates of the prevailing scientific theory are sometimes so doctrinaire that they ridicule one who dares to depart too far from the prevailing mode.   James dared such ridicule.   Above all, he had that childlike spirit that continued to grow and develop new ideas to the very end—new ideas that rendered his earlier theories untenable.

At the last William James hit upon his radical empiricism. We believe, if he had lived long enough to recast his earlier thought in the light of this latest view, he would have revised statements in *Varieties of Religious Experience,* such as the following:

> Psychology and religion are thus in perfect harmony up to this point, since both admit that there are forces seemingly outside of the conscious individual that bring redemption to his life. Nevertheless psychology, defining these forces as "subconscious," and speaking of their effects as due to "incubation," or "cerebration," implies that they do not transcend the individual's personality; and herein she diverges from Christian theology, which insists that they are direct supernatural operations of the Deity. I propose to you that we do not yet consider this divergence final, but leave the question for a while in abeyance—continued inquiry may enable us to get rid of some of the apparent discord.[4]

We believe that his radical empiricism, consistently applied to interpretation of religious experience, would have brought him much closer to Christian theology than this. But let us see how his "continued inquiry" gets "rid of some of the apparent discord." On page 242 he says:

---

4 p. 211.

But if you, being orthodox Christians, ask me as a psychologist, whether the reference of a phenomenon to a subliminal self does not exclude the notion of direct presence of the Deity altogether, I have to say frankly that as a psychologist I do not see why it necessarily should. The lower manifestations of the Subliminal, indeed, fall within the resources of the personal subject: his ordinary sense-material, inattentively taken in and subconsciously remembered and combined, will account for all his usual automatisms. But just as our primary wide-awake consciousness throws open our senses to the touch of things material, so it is logically conceivable that if *there be* higher spiritual agencies that can directly touch us, the psychological condition of their doing so *might be* our possession of a subconscious region which alone should yield access to them. The hubbub of the waking life might close a door which in the dreamy Subliminal might remain ajar or open.

This suggestion of James has been seized upon by many religious people as a refuge from all the disturbances of scientific investigation. Here, they have felt, we have a place to rest our religious convictions which is inaccessible to all the currents of scientific thought and prying psychology. *If there be* a sublimal mind wholly inaccessible to all prying psychologists, so deep down in the self that it cannot be investigated, and *if there be* higher spiritual agencies, then perhaps in this deep unknown and unknowable region of ourselves we stand face to face with God, laved and embalmed by his constant and intimate presence. The only trouble with this theory is that it reconciles science and religion as Herbert Spencer did with his Unknowable and as Kant did with his region of the noumenon. To be sure any assertion one cares to make about that region of the subliminal self cannot be refuted. Neither can it be proven. Any belief one wishes to cherish is beyond the reach of all experimentation. It cannot be put to the test of experience. It cannot be scientifically investigated because it lies outside the field of accessible experience. But of what use to religion is a belief that cannot be put to the test, that can be denied as readily as affirmed without any check in experience to distinguish truth from error? James would say that there is one check. If one belief helps you, makes you happy, gives you courage and zeal and hope and power, and the opposite does none of these things, and if both these beliefs refer to a region that is wholly beyond all experimental verification or investigation of any sort, then plainly one is a fool not to accept the helpful belief, especially when the stake at issue between the two beliefs may be a matter of life and death to the human race and success or failure in the life of the individual.

But we claim that just as science has no use for experience which is wholly inaccessible to awareness, neither has religion. It is true that there are some who cultivate religion because it unleashes their phantasy; because it enables them to float away into a cuckoo-land of dreams unchecked by any hard, grim facts. Such persons cultivate religion for the same reason that others read fantastic fiction or take drugs or go on drunken sprees—it enables them to forget reality and comfort themselves with the free play of phantasy. We do not for a moment deny that such religion prevails quite widely. But for that matter there are also amateur "scientists" who cultivate "science" in the same way and for the same purpose; and there are dilettante farmers who cultivate farms, and so on throughout all the walks of life. The dilettante in religion is not peculiar to religion. We find him everywhere, always seeking some region where he can dream undisturbed by disagreeable facts and where his phantasy can rear its castles in the air. But it is just as false and unfair to take the religion of the dilettante as a true instance of religion as it is to take the science or the love or the farming of the dilettante as true instances of these several human interests.

Religion claims to deal with fact, with objects of immediate experience, just as much as science, just as much as engineering or farming or any other of the major concerns of human life by which men live. The earnest upholders and propagators of religion, in our time and in all time, have claimed to deal with that which shattered false dreams and thrust itself into human affairs with all the inevitableness and unyieldingness of ultimate fact. And if it could have been proven to these earnest upholders and propagators that religion really did lie beyond the reach of experimental testing and immediate experience, they would have had nothing of it. They have been men and women who craved exposure to fact and first-hand experience of that which is truly existent.

But let us consider further what James says in that very significant statement which we last quoted. He says that "the ordinary sense material, inattentively taken in and subconsciously remembered and combined, will account for all his usual automatisms." Here he notes a supremely important distinction but does not seem to make further use of it. It is the distinction we have been insisting upon between the sense material, the datum, the experience of the not-self, and those automatisms or other responses of the self to such experience. We all know James well enough to be sure that he would not claim this "sense material inattentively taken in" to be created by the self. It is given to the self from the outside. It is the most direct access which the self has to the external world. To experience this sense material is not to experience the subliminal self merely.

But there is another thought, even more important, in these words of James. He says that "Wide-awake consciousness throws open our senses to the touch of things material," but the subconscious, "the dreamy Subliminal," may yield us access to the spiritual. Here James has touched the central issue. Here he has touched the sore spot that is causing all the trouble. Right here is the parting of the ways, one leading with false hopes into endless confusion and recurring scepticism, the other, while more forbidding at the start, leading to the only certain knowledge there can be about things spiritual. James suggests that "if there be higher spiritual agencies" they cannot be found in the world of sense. Things material must exclude things spiritual; we must turn away from the material world, close all the senses that yield us knowledge of it, and find some hidden other sense which will give us knowledge of the spiritual world. Here is that pitiful blunder that always leads to confusion. Here he has put his feet to that path that leads out into the morass where nothing but dreams and will-o'-the-wisps can be found. Turn from the wide-awake consciousness of sense, which gives us knowledge of the material world, and turn to the consciousness of sleep, turn to dreams, to "the Dreamy Subliminal" and perhaps there we can find the spiritual. As sure as any one turns down that way, clear-headed thinkers sooner or later will show him to be following illusions, phantasies, and myths. James did not himself go that way. But with that broad sympathy and marvelous acquaintance with all the ways of the human heart, he recognized this to be the way so dear to many, and, in passing, acknowledged it to be a way which some might wish to follow; and he would not close the path to them.

But we must insist that if the spiritual is to be found at all it must be found in and through the material. The same senses that reveal the material must also reveal the spiritual. And in fact is that not very plainly the way in which we become cognizant of, say, other human minds which are spiritual entities, if the word spiritual has any significance at all. We do not mean to assert that the material is anything else than material. Neither do we mean that everything has both a material aspect and a spiritual aspect. There are material things which are wholly material and have nothing spiritual about them. In the same way we assert that oxygen is nothing but oxygen. It is not oxygen in one aspect, and water in another. But oxygen, when combined with other elements, makes water. So also material elements may be organized in such a way as to constitute the spiritual. And that does not make the spiritual dependent upon the material, but quite the contrary. When the material has been organized into a spiritual being, that being may be able to maintain itself. In fact, one unique character of mind, is that it resists disintegration with more energy and ability than does matter which has not been integrated into a mind. And what is most important, the whole material world, while not ceasing to be material, may in its totality, by reason of the form of this totality, constitute a mind. This problem of matter, mind and God, we cannot here discuss at length.[5]

But let us turn to another statement of James. With reference to his study of the mystic experience ensuing from the use of certain drugs, he says:

> One conclusion was forced upon my mind at that time, and my impression of its truth has ever since remained unshaken. It is that our normal

[5] Whitehead, A. N., *Science and the Modern World.*

waking consciousness, rational consciousness as we call it, is but one special type of consciousness, whilst all about it, parted from it by the filmiest screens, there lie potential forms of consciousness entirely different. We may go through life without suspecting their existence; but apply the requisite stimulus, and at a touch they are there in all their completeness, definite types of mentality which probably somewhere have their field of application and adaptation. No account of the universe in its totality can be final which leaves these other forms of consciousness quite disregarded. How to regard them is the question—for they are so discontinuous with ordinary consciousness. Yet they may determine attitudes though they cannot furnish formulas and open a region though they fail to give a map. At any rate, they forbid a premature closing of our accounts with reality. Looking back on my own experience, they all converge towards a kind of insight to which I cannot help ascribing some metaphysical significance. It is as if the opposites of the world, whose contradictoriness and conflict make all our difficulties and troubles, were melted into unity. Not only they, as contrasted species, belong to one and the same genus, but one of the species, the nobler and better one, is itself the genus, and so soaks up and absorbs its opposite into itself.[5]

How wonderfully close James here comes to what we believe to be the truth. That noble, generous soul so eagerly seeking the truth, following

[5] *op. cit.* p. 388.

every clue with eyes wide open, comes so near the door. Just a step more and he would be inside. We are tempted to cry the quotation which he himself uses: "Oh, the little more, and how much it is; and the little less, and what worlds away." And the wonderful thing is that he himself has given us the key to open the door which he could not enter. It is no honor to us that we can go where he could not. It is only because of him that we can enter in. That key of radical empiricism which he found so late in life, too late to use himself for the opening of many doors, he has given to us.

There are "other forms of consciousness," he says, which we ordinarily disregard, different from "our rational consciousness." That there are these other forms of consciousness there can be no doubt. The only question at issue is "how to regard them . . . they are so discontinuous with ordinary consciousness." Yes indeed, here is the question at issue. James could get nowhere with the problem as long as he considered these "other forms of consciousness" as discontinuous with our normal states of consciousness. They are not discontinuous. They are directly continuous.

The only difference between them and normal consciousness is that in them we become aware of a mass of merged data, so merged as to be a single unanalyzed and unsifted datum, while in ordinary consciousness we are aware only of a very few data which have been selected from out of this flood. Here is perfect continuity. And these "other forms of consciousness" are not different from the rational, if by that one means they are wholly irrational, or forms of consciousness to which rationality does not apply.   If by rationality one means the analysis and selection of data to the end of discovering just what each datum or group of data signify, then these "other forms of consciousness" provide data which are just as much subject to interpretation as the data of any other form of consciousness.   If by rational consciousness one means consciousness in which the data of awareness have been more or less completely and correctly interpreted, then of course the mystic experience is not rational.   That is to say, it has not been interpreted but it is subject to interpretation.   If by notrational one means a form of consciousness which is not subject to rational interpretation, then these other forms are not not-rational.   They are forms of consciousness in which we are aware of the same world in which we live constantly.   They are simply states in which our awareness is more diffusive, less selective and analytic.

CHAPTER XII

RELIGION AND REFLECTIVE THINKING

It will be impossible for religious thought to escape the influence of John Dewey's latest, and perhaps greatest, work, the Paul Carus lectures published under the title of *Experience and Nature*. He makes very little explicit reference to religion but his ideas have important bearings upon religion. His thought is one of the noteworthy forces shaping modern life and anything so pervasive as religion cannot escape its touch. There are two ideas running through his recent work which we want to develop and use as searchlights to illumine the nature and function of religion.

The first of these is his concept of meaningless experience. The prevailing philosophic tradition has declared that all experience is meaning and without meaning there can be no experience. But Professor Dewey makes a sharp distinction between meanings and mere events which occur in space-time, whether these appear in the consciousness of some individual or not. It is quite possible for experience to occur as a meaningless event. Meaning is the significance which some event or events may have. Significance is the pointing function, the signifying, which some events may possess.

But in so far as events may occur in one's experience without signifying anything to the individual, they are meaningless for him. Such experiential events do occur, he claims.

We shall make use of this concept of meaningless experience to interpret mysticism. Mysticism is meaningless experience when the latter is believed to be an intimate association with God. Of course the natural rejoinder to such a remark is: But it ceases to be meaningless when it is viewed as association with God or valued in any way. But we must distinguish between beliefs about an event, and the meaning which the event itself may have for the individual during his experience of it. If he does not think during the experience, the experience is for him at the time meaningless. The mystic state is not a thinking state. Of course thinking may occur in all degrees, but it approaches extinction in mysticism and may disappear altogether.

That mysticism is a state in which experience attains richest concrete fullness of content with minimum of meaning, is the view of it upheld by the two foremost expounders of mysticism in the United States, Professors Hocking and Leuba.[1] That these two writers should differ so radically in their total evaluation of mysticism and yet should agree perfectly on this one fundamental issue, is striking confirmation of our thesis concerning its nature as meaningless experience.

[1] William Ernest Hocking, *The Meaning of God in Human Experience*, and more recently an article on "Principles and Methods in the Philosophy of Religion," *Revue de Metaphysique et de Morale*, XXIX, 1922, 431–53. James H. Leuba, *The Psychology of Religious Mysticism*, particularly p. 313.

It will be our endeavor to show the value of this practice of mysticism which is worship and which involves casting off all our old meanings. Its value is (1) that it enables us at times to develop radically new meanings; or (2) to come back to the old meanings with new freshness and vigor; (3) to free ourselves for a time from the binding tension and constraint of established meanings imposed upon us; (4) to quicken our sense of the concrete fullness of experience underlying our meanings by dipping into that stream of total event to which all our meanings must ultimately refer if they are to be efficacious in controlling the conditions of life.

To demonstrate the value of this mystic meaningless experience we shall draw upon no other than Professor Dewey himself, although he is not ostensibly an expounder of mysticism. There are many forms and motives in mysticism; and what Professor Dewey would condemn we also perhaps would condemn. But there is a form and use of mysticism which Professor Dewey may not condemn and which, in any case, we are very sure is of the highest service in promoting the goods of life. This form and use of mysticism reveals, we believe, one of the high and indispensable functions of religion.

The second searchlight we want to borrow from Professor Dewey to illuminate the nature and value of religion is his concept of meaning. Meaning is the mental instrument by which we control experience and magnify its value. Life mounts in value and security just in so far as we bring into operation the right meanings and use them rightly. The whole problem of human life, in a nutshell, is just this: How to magnify the values and security of life. This problem finds its solution in meaning providing meanings can be torn down, reconstructed and progressively elaborated by the radical method of mysticism. A set of radically new meanings can arise only when old meanings are discarded; and between this discarding of the old meanings and the rise of the new, there is an intervening state of consciousness which is relatively meaningless. This is the state of mysticism. This is the state of all profound worship. The bringing on of this creative and regenerative process is one of the supreme functions of religion.

The order of our procedure shall be, first, to clarify the concept of meaningless experience; second, do the same for the concept of meaning; and, third, apply these concepts to the interpretation of mysticism.

MEANINGLESS EXPERIENCE

Ordinarily human experience is not meaningless. It may be questioned whether experience can ever attain consciousness without some rudiments or vestiges of meaning in it. We believe it can, and that this occurs much more widely than is ordinarily thought; but whether or not that is granted, we hope to show that meaning can be reduced to a minimum without proportionate diminution of consciousness.

Undergoing some excruciating pain, where the mind is reduced to a blur of agony, is a case of meaningless content of experience. The flood of emotion that pours over one in response to music is a meaningless experience providing one does not attend to the music, think about it or discriminate its several qualities, but simply yields to the emotional state which some music is able to generate in some people under proper conditions. Basking stupidly in the sunshine is another instance. The enjoyment of a warm bath providing one does not cognize the water nor anything else but simply submits himself to the voluptuous experience of that which is occurring then and there. Sensuous enjoyment of good tasting food without any recognition of what it is we are enjoying is a case in point. That expansive, beaming state of well-being which some healthy human animals display when they come away from a meal, seems to be of this sort.

Of course one can well insist that an adequate appreciation of music or food or any other object requires a finely developed system of judgments by which to discriminate and comprehend all the qualities that enter into it. With that we would thoroughly agree. But we are not talking about what is or what is not adequate appreciation. We are simply trying to point to cases where concrete content of experience occurs with little or no meaning.

The difficulty of pointing out or calling to mind any case of meaningless experience is that such experiences leave no marks behind them. Only meanings endure. Content without meaning passes like a baby's breath. Qualities that merely occur, but are not discriminated and have no meaning for the person who experiences them, cannot be kept in mind. And they never recur; they pass beyond recall, *spurlos versunken*. It may be that the instances we have mentioned could be called to mind only because they contained some rudiment or vestige of meaning. But if we can trace content of experience to the point where meaning fades to the minimum, we have established all that is required of us. We can catch experience at those moments where all old meaning is just about to fade out completely, or where new meaning is just coming into existence.

There are certain abnormal states of mind where consciousness can be very vivid and rich but with scarcely any meaning. Epilepsy may assume this form, and certain drugs produce this effect. Professor Leuba has made a very thorough and finely analytic study of many cases of this "mystic state." He begins with the description of the effects of certain drugs such as mescal, hashish, stramonium, alcohol, or of gases like ether and nitrous oxide. He notes how some savages have used these drugs to bring on that mystic state which was in their view a kind of divine possession. Then he notes other artificial methods of bringing on this meaningless content of consciousness by rhythmic movements carried to a frenzy of dancing, or again by long fasting or self-tortures, or a combination of these. By these methods primitive folk produce a delirium of consciousness which is rich and vivid but meaningless. The Yoga system of mental concentration is a much more refined method of producing the same result. Most of these generally end in the fading out of consciousness altogether, but prior to the elimination of consciousness there is a state where conscious content persists but meaning has disappeared. The great Christian mystics, according to Professor Leuba, have simply carried the same practice to a much higher level of refinement and turned it to ethical ends but have not changed the essential nature of the experience.

Before leaving this subject some explanation should be made of the fact that most moderns, and especially thinkers and philosophers, have so commonly ignored this sort of experience, sometimes even denying that it ever occurs in the normal human being, and generally failing quite completely to recognize the important function it has in the promotion of more abundant life.

Certain philosophers in particular have been most insistent in claiming that all experience must be meaningful, that there is indeed nothing else in the universe except meaning. That thinkers should have this bias is easily understood. They are preoccupied with meanings to the exclusion of all else, precisely because they are thinkers. Add to this natural bias the weight of successive generations of thinkers who gradually shove their followers farther and farther into the realm of meaning to the ignoring of any other ingredient in experience. Finally the student of philosophy must perforce immerse himself for years in the study of ponderous and technical works dealing altogether with abstract meanings which for the most part have no apparent reference to the immediate events of space-time. Considering all this it is no wonder philosophers have been so commonly prone to swallow up all experience in meaning.

But the ordinary man is often subject to the same fallacy. We have already noted the difficulty of recalling or referring to any experience bereft of meaning. Meaningless experiences can be enjoyed, but only meaningless ones can be used. The ordinary man fixes his attention generally on that which he can use. A further prejudice against meaningless experience arises from the feeling that it is not befitting the dignity of *homo sapiens* to fall into that bovine stupidity where consciousness has no meaning. For the sake of our natural conceit and self-respect we will not admit that we ever yield ourselves up to that enjoyment of the eating situation where all discrimination and meaning fades out into meaningless satisfaction. I admit the premise that it is not nice to do this; but I do not admit the second premise that humans are always nice. It is not considered quite so disreputable to yield one's self to the purely effecto-motor state of enjoying music in which floods of emotion without meaning pour over one. But even in such cases people are likely to claim that they contemplate some profound unwordable truth or have great insights revealed to them through the medium of music. That such experiences may provide the way to new insights and great truths, we do not deny. We only assert at this point that such experiences may be, and often are, quite meaningless and shall further show that it is just this lack of meaning in them which makes it possible for them to serve, at times, as transitional stages toward the development of new meanings.

To guard against another possible misunderstanding, we must point out that we are not asserting all music must be enjoyed without meaning. There may be no content of experience whatsoever which does not have its proper meaning if it were known. But what we do claim is that music and food and many other goods may be enjoyed by a human without grasping any meaning in the experience; and that this is much more common than self-complacent mortals ordinarily admit.

MEANING

Meaning, otherwise called the concept or judgment, is the method by which we control the content of experience. Meaning first arises through a relation that is established between some sound or other gesture, and some event. The sound or gesture then becomes a symbol with a meaning. The meaning is not related to the event. On the contrary the meaning is the relation of the gesture to the event. We do not first have meanings that must later be related, as Bradley would have us believe. The hopeless difficulties into which we fall when we follow his lead in that direction is a pitiful spectacle of what trouble philosophy may cause itself. This error of trying to relate meanings to events or to other meanings, as though the meaning was something else than precisely that relationship itself, could never have arisen had not meanings become divorced from events after the fashion we shall shortly describe. But, as we have said, when a sound or gesture becomes so related to an event as to mean that event, it becomes a symbol and its meaning or significance is precisely that relation to that event.

But meanings point in two directions, toward events and toward other meanings. When a gesture such as sound, or movement, or mark, enters into that relation to some other event by virtue of which it has meaning, and some other gesture likewise acquires meaning in the same way, these two meanings may enter into a meaningful relation to one another. Thus a meaning may mean not only an event but also another meaning. In fact, these two lines of development proceed side by side. As a gesture refers to an event so also it refers to other meaningful gestures. And its meaning with respect to the event can become clear and definite only as it comes to involve other meanings. In other words, it requires a whole system of meanings adequately to define an event. Meaning becomes more meaningful just in so far as it specifies more accurately and completely some event or class of events, but this it can do only as it becomes elaborated into other meanings. A meaning is like a vulture which, as soon as it begins to hover about an object draws other vultures to it.

But right here a most amazing twist of affairs often enters in. This relay race of one meaning passing over into another and the other into still another and so on, may become so interesting that thinkers become absorbed in it to the exclusion of all events whatsoever save those meager symbolic sounds and marks which serve as vehicles of meaning. This is that divorce between meanings and events to which we referred above. Not only are professional thinkers often caught in this whirl of abstract meanings because of their love for mental gymnastics, but ordinary men also may find their well-worn meanings so sufficient for their immediate needs as to ignore to the utmost any events to which these meanings might refer. Thus we often get on the high stilts of meaning and never touch the ground of events. But the whole great manifold of events or, what is the same thing, the Total Event, is going on all the time just the same, whether we have any sense of it or not. And this manifold movement of time-space will sooner or later break down our little structure of meanings and destroy us and them if we do not constantly reconstruct and renew our meanings in such a way as to discriminate and correlate events.

Professor Dewey[2] illustrates meaning by the significance of a policeman's whistle at the street corner of congested traffic. His whistle means, first of all, a certain specifiable behavior of the traffic. But these specifications for traffic may be considered, and often are considered, quite apart from any

2 *loc. cit.*, p. 196.

particular instance of traffic behavior. Furthermore, these specifications require and imply still other specifications and regulations. Ultimately they involve that whole system of specifications which make up the law and government of the country. These specifications are not in themselves physical, nor are they necessarily mental. Of course they become mental when a thinker thinks them. But they are constituents of the total universe even when no thinker happens to be thinking them. They are methods by which we control certain instances of experience in so far as experience is involved in the social behavior of that group which is subject to these regulations. These specifications render meaningful any particular experience that comes under their control. Any event which occurs to me means much or little according to the scope and fulness of the specifications which control it.

Of course such specifications or meanings may be of an entirely different order from that of the policeman's whistle. They may be the specifications of pure mathematics. The whole science of mathematics is such a system of specifications, some of which apply to actual events and some of which seem to have no reference to any possible events whatsoever.

Now these specifications, whether of mathematics or of the governmental system, may become matters of such absorbing interest to the mathematician or the jurist that they lose sight of all events to which the mathematical or legal system originally referred. Not only so, but the man who is neither mathematician nor jurist, who is not a thinker at all, but a stupid routineer, may blindly conform to the specifications of some mathematical or legal system without regard for any events to which they are applicable. The practical evils that result are matters of common knowledge. The "legal mind" has become a by-word. The mathematical theorist whose formulas are flawless but will "not work" is well known. But the so-called practical evils are not the only ones, nor are they necessarily the most important. This process by which meaning becomes divorced from events is one in which meaning commits suicide and all the spiritual values of meaning slowly die out. But before we discuss this matter let us approach the matter of meaning from another angle.

Meaning is the method by which we control experience. Now we control experience by means of movements. Hence the first thing meaning must do, if it is to serve effectively in controlling experience, is to specify the motions that enter into the experience, *i.e.*, the space-time relations. The events of experience are in space-time, hence the meaning that controls them must indicate accurately these space-time relations. Hence scientific analysis of experience reduces very largely to such a system of relations. The meaning of color, for physics, is a system of vibrations, which are space-time relations. Sound, heat, hardness, weight fluidity, are reduced by physics to space-time relations. The difference between gold and silver, between helium and uranium, are, in terms of physics, due to difference in space-time relations. Meaning, then, is largely reducible to these because content of experience can be controlled only in such terms.

But there is something more than this in our meanings which are our system of judgments. In order to control experience we must be able to discriminate the different qualities that enter into it as well as trace the space-time relations in which they occur. In order thus to discriminate we must be able to point out one element of a total situation and distinguish it from the others. Thus judgment or meaning must consist of a very elaborate and delicate system of pointers. The word yellow, for instance, is a pointer by which we distinguish certain occurrences in our experience from other occurrences. No two such occurrences are ever the same; but they are all of such a nature that they can be grouped together and so designated all together or singly as yellow. The reason why they can be grouped together in this way is because the pointer called yellow, when put into operation, is so designed as to fall upon these particular occurrences and no others.

Of course the word yellow, taken merely as a word, is not a pointer. There must be a system of judgments back of it to give point to it. It could be compared to the apex of a pyramid. The extreme apex of a pyramid, when separated from the pyramid, is not a point at all. It is a mere speck floating about and indicating nothing. The pyramid or judgments back of the word are what give indicative value to it.

But pointers of such a nature must be constructed and operated according to certain rules. Otherwise they will not point. These rules for the construction and operation of pointers constitute what we call logic in the strict and narrow sense. Inconsistency in the system of judgments, for instance, destroys the indicative value of the pointer. You do not know what it points to because it seems to point to two or more wholly different things; or else it points to nothing at all because the judgments do not constitute a system at all. The pyramid is broken up into fragments with no apex or with many different apexes.

Now we have noted how much a system of judgments, constituting a pointer or system of pointers, may engage our care and attention to such a degree that we lose sight of that content of experience to which the system points. Mathematics may evolve a system of meanings which have no reference to any content of experience beyond themselves, but are cultivated for the aesthetic enjoyment of the system itself. Of course it is an excellent thing to have on hand such elaborate systems of meanings in case they may be found to be applicable to the control of events at some future time, even though they seem wholly inapplicable at the present time. Such development of useless meanings that have later been found useful, is an old story in science. We are only pointing to the way in which a system of meanings may, and particularly in philosophy frequently do, become content of aesthetic enjoyment pure and simple and cease to function, as meanings properly should, to control the rest of life. When meanings do this, they cease to be meanings in the fullest sense; they become content rather than meaning. That meaning should cease to be meaning may seem to be paradoxical. But if we follow closely our definitions I think the paradox disappears. Meaning is that which refers, on the one hand, to certain events, and on the other, to other meanings. When it fails to do either of these it becomes crippled as meaning; it becomes incomplete as meaning. When it ceases to have this dual reference, it loses some of the values of meaning.

When meanings turn back upon themselves and become a closed system referring to nothing on beyond themselves we have what Plato and Aristotle seemed to think was the noblest, the most real, phase of the universe. Many think this false estimate was the greatest weakness of Greek thought. Certainly the inheritance of this bias down through all ages of thought since then has been a great incubus. According to Aristotle, God spent all his blissful time engaged in such purely aesthetic and meaningless operations of the mind—meaningless in the sense that the judgments indicated nothing in the field of events that occur in space-time and had no application there, hence could not be used to control experience or magnify and secure the common goods of living. Such view of truth as a closed system of concepts referring to nothing beyond itself is probably the reason the Greeks could so easily identify truth and beauty, for such a system is certainly an aesthetic object.

There has been a great deal of philosophic thinking that has come to such an impasse. It has lost all bearing on events of space-time. Philosophers who develop such systems naturally are prone to deny any distinction between content of experience and meaning.

There are, then, two ways in which content of experience may become meaningless. One is that events of experience should occur prior to the development of any relevant system of judgments. The other way in which content of experience may become meaningless is through the development of a closed system of judgments of such nature that they do not control or point to anything beyond themselves. Such a system has no meaning or significance because significance is the function of pointing or signifying and such a system does not do this, except as between its own constituent members. Of course systems of judgment are rarely closed. Even when they are developed in such a way as to lose their function of signifying and controlling events, they may still point on to further meanings. Such a system is not altogether meaningless, but it has lost that original function of meaning which consists in controlling events. It is neither true nor false so far as events are concerned, because it does not apply to these. It has lost at least a part of its capacity to render life more abundant.

It is plain that a great transformation occurs in life, and in the universe with which life deals, as soon as meaning begins to engage attention and determine conduct, providing meaning maintains its dual reference to events and to other meanings. When meaning of this sort flings its rainbow over the earth, spiritual life begins. It is through the development of symbolized meaning that science, art, and morality can come into existence. It is through such meaning that sex becomes love, the biological adjustments of maternal care become affection, the herd becomes the state, signaling becomes discussion, food-getting becomes industrial organization, etc. It is through such meaning that states of consciousness can become indicative of a wide complex system of behavior and experience which far transcend the time and place of their occurrence.

The deepest drive of human life is to render itself more abundant. To become more abundant means to have access to wider ranges of experience for use and enjoyment. The one supreme and indispensable means to this increase of life is meaning. But meaning may fail, and so life may fail, in either of two ways, apart from the failure due to false meaning. Meaning may fail to promote life through the loss of all meaning whatsoever, or it may fail by the development of meanings that no longer apply to the events of space-time. In the one case meaning dies, leaving event a widow; in the other case meaning becomes divorced, leaving event a divorcee. The one is the evil of hyper-sensuality; the other the evil of hyper-rationality.

Of these two dangers to which life is constantly subject, one the occurrence of events which have lost their meaning or never acquired any, the other the operation of meanings divorced from events, we believe the latter is by all odds the greater at our present stage of civilization. Of course the two evils often go together. But as civilization develops and becomes more complex this danger seems always to increase. The complexities of meaning which control our lives become so great, and absorb so much of our attention, that we are in danger of ignoring the concrete vivid fulness of qualities that occur in experience. One result of this is that when these qualities of space-time change beyond a certain limit, as they always are changing, we will fail to note that our system of meanings no longer applies and hence no longer serves adequately to control and direct our lives. Outworn laws, obsolete institutions and methods, are constant reminders of this danger. When maladjustment goes too far, disaster ensues. But even when this disaster does not occur, when the meanings serve very well to keep us safe in the midst of a tumultuous and dangerous universe, still our lives are pitifully impoverished if they become wholly subject to meanings which do not illuminate for us the qualities of space-time that occur in our experience. When the meanings which regulate our lives do not make us more vividly and widely aware of hands and smiles and glancing eyes and wind and cloud and dust and flower, our meanings are not properly functioning.

One of the great values of mysticism, as we shall see, is that it enables us to escape from such a hard outer shell of meaning. The mystic experience is a meaningless, conscious event. In mysticism we discard all our old meanings and consciously submerge ourselves in the total event of experience. Out of such an experience we may emerge in a mental state which develops new meanings or modifies and reinterprets the old meanings and readapts them to the events of space-time. This brings us immediately to the subject of mysticism.

MYSTICISM

We have hinted that mysticism and radical originality may be closely allied. Mysticism does not necessarily lead to originality, any more than eating food necessarily leads to renewed strength and energy. Often just the contrary occurs; the mystic may of all people be least original. But the mystic experience is one which makes originality possible. Here again we get help from Professor Dewey.

Professor Dewey notes two kinds of originality. One consists in organizing and reorganizing established meanings in such a way as to bring forth new meanings. But the other is far more difficult, more rare, and far more fruitful in its possibilities of enriching the meaning of life. This other method is that of initiating new methods of viewing and dealing with the raw materials of experience. The first type of originality consisted in building on to the old meanings. The second type consists in going behind the old meanings and constructing new meanings with respect to the event of immediate experience. In this latter type the old meanings are resolved into something different, not merely carried on to their further implications.

Following is the way Dewey himself states the contrast:

> There is a difference in kind between the thought which manipulates received objects and essences to derive new ones from their relations and implications, and the thought which generates a new method of observing and classifying them. It is like the difference between readjusting the parts of a wagon to make it more efficient, and the invention of the steam locomotive. One is formal and additive; the other is qualitative and transformative.[3]

Then he goes on to describe the mental state in which this radical sort of originality occurs:

> When an old essence or meaning is in process of dissolution and a new one has not taken shape even as a hypothetical scheme, the intervening existence is too fluid and formless for publication even to one's self. Its very existence is ceaseless transformation. Limits from which, and to which, are objective, generic, stateable; not so that which occurs between these limits. This process of flux and ineffability is intrinsic to any thought which is subjective and private. It marks "consciousness" as bare event.[4]

Now this "consciousness as bare event" is precisely that form assumed by immediate experience

---

[3] loc. cit, p. 222.
[4] loc. cit., p. 221.

when there is no meaning in it, or when meaning is at the minimum. It is experience caught in that intervening period when old meanings have faded out and new meanings have not been born. It is experience prior to the discovery of meaning. It is experience which we feel but do not think, although in the case mentioned one is struggling to think. But he is struggling to think in some new way, not in the old way. When one goes directly to consciousness as bare event and struggles to develop meaning out of it, one has the experience described by Professor Dewey.

Now this experience, we claim, is identical with religious mysticism at its best. It reveals, furthermore, the vital indispensable function in human living which religion has to play. It is the regenerative, recreative function. This experience which Dewey describes, which manifestly is an experience he himself has undergone, for his originality is of this radical sort, this experience which is consciousness as bare event struggling with unborn meanings, this is the mystical experience par excellence. The mystic at his best is the midwife struggling with immediate experience to bring new meanings to birth.

It is true that this state of consciousness as bare event may be, and often has been, cultivated by the mystics for its own sake and not for the sake of the unborn meanings in it. They have luxuriated in the experience without any attempt to bring forth new meanings. But this misuse of mysticism does not militate against the proper use and value of it any more than the misuse of gustatory experience condemns eating.

This state of consciousness described by Professor Dewey is by no means to be identified with that of the lower animals or the primitive savage, even though it be an experience bereft of meanings. We have no reason to think that the consciousness of animal and savage consists of any awareness of such depth and fullness of immediate data as enters into the "fluid and formless" flux of the radically original thinker during that intervening period after the old meanings have faded out and before the new meanings have emerged. On the contrary we have every reason to think that the primitive and the animal have the most impoverished content of consciousness. Experience can enter consciousness with any large fullness only as it is illumined by meanings. But meanings, when they attain any high degree of systematic completion and fixity, do not illuminate, but rather veil, the data of immediate experience. It is when meanings are being born or undergoing reconstruction that they quicken to the maximum the consciousness of concrete experience. In that "process of dissolution" described by Professor Dewey, it is the old meanings just fading out, and the new meanings just being born, that make consciousness so rich and fluid in content. It is because experience, while bereft of any developed meaning, is pregnant with meaning, that its depth and richness can engage attention. This plainly is not the experience of the lower animals and the primitive savage.

It is the saving function of religion in human life to foster this mystic consciousness which is a condition of radical originality. Worship at its best is precisely this. It is the great regenerator, renewer, and reconstructor of human life because it fosters that experience which provides for the extreme reconstruction of meanings. It revitalizes old meanings with new insight, brings on "conversion," and once in a while it lifts human history bodily into new channels, as shown in those periods when great religions have been born.

At the greatest turning points in history we find a mystic. Whether Jesus or Paul or both be considered founders of Christianity, we have in Christianity one of the most tremendous original historical achievements of history issuing forth from the mystic. Of course all the ingredients that enter into Christianity may be traced back to earlier sources. That is true of every historical phenomenon. But in the sense that any historic achievement was ever original, Christianity was. So also with other turning points in history. Buddha, Mahomet, St. Francis of Assisi, each broke into long stretches of uniform history with a transforming originality.

Of course many a man may have access to this experience of the mystic and not have the constructive intellectual powers to develop new meanings except in a very vague and rather futile way. But even these vague and inchoate meanings have some suggestive value. And when a keen intellect is also somewhat of a mystic, we have the source of the most valuable kind of originality, such as Dewey himself exemplifies.

That mysticism should issue in radical originality has its ready psychological explanation. In the mystic experience one becomes freed of all his old established meanings; he is lifted out of the ruts of all his ordinary thinking; he is shaken loose from all his mental habits; his mental system is melted down into a fluid state; he becomes filled with a throng of free, uncontrolled impulses so numerous and diverse that they hold one another in abeyance, producing a quivering mass of sensitivity to the total undiscriminated situation. When his mind recrystallizes out of this state it is quite likely to assume a different form from what it had before the experience occurred. The reflective thinker approximates this mystic deliquescence of meanings in his endeavor to deliver himself from all bias of prejudice and habit and give free play to every impulse in order to view the problem from every angle. However the two may differ in the motives that lead them to this state of mental deliquescence, the thinker and the mystic at this extreme point do truly enter into the same general type of experience.

And what is more, they do not always differ so greatly in motive. There are all sorts of mystics, but the best of them, and most of them in their best periods, have entered the mystic state not merely to luxuriate in the deliverance from all constraint of habit, and the free play of impulse, and the social esteem which came to them because of the supposed divine visitation involved in this experience, and other such considerations, but they entered it in their struggle to solve problems. They have almost all been individuals who were struggling with the most serious difficulties of adjustment and undergoing most severe conflict of impulses. It was their effort to find a more satisfactory way of life and so solve the ultimate problem of all human living that gave rise to this state. They were searching to find the solution of a problem and the greatest of them, as well as many who have attained no fame, ultimately found the object of their search, namely, a unified, harmonious and effective way of living.

The mystic considers his experience to be a form of worship and communion with God. This also is not so remote from the reflective thinker as some would believe. The reason it seems so remote to many is because the kind of reflective thinking we are here considering, which leads to an experience approximating that of the mystic, is so very rare. Most thinking is not original, and of that which is original by far the greater portion of it is of that first type of originality mentioned by Dewey—the further elaboration of old meanings rather than the introduction of a new method and outlook. Most thinking is but clever manipulation of old meanings, defense of established positions, or deduction of implications. But the kind of original thinking which leads to the "mystic" experience is one in which the thinker struggles to divest himself of every bias and limitation imposed upon him by his mental habits and established meanings. He struggles to get away from himself, understanding by self his established system of meanings. This struggle to escape from self with all its limitations and prejudices, this profound effort to open one's mind to the total fact, is very closely akin to worship. It is somewhat of the same motive, the same earnestness and profundity, the same method, as the worship of the mystic.

Yet one cannot think without a system of meanings. How can one escape from the only system he has and at the same time have a system for the purpose of thinking? Only by disrupting his established system into a wide free play of impulse and allowing it to recrystallize. This is what the thinker does and this is what we have seen the mystic does. Yet the thinker does not lose all meaning and direction in his experience and neither does the mystic. He enters the experience with a purpose—to get a more adequate system of meanings—and this purpose gives direction to the whole process. One difference between the thinker and the mystic is the degree of control that still persists throughout the experience, it being presumably greater in case of the thinker than in case of the mystic. And yet here also the difference is too commonly magnified. As Dewey insists, the thinker cannot forecast the outcome. He must be ready to relinquish every precious object of desire, every good that has been cherished, for out of the process of radical originality may come that which is strange and unloved heretofore while all that was dear is cast away.

Of course the chief difference between the mystic and the thinker arises from the difference in their beliefs about the nature of the experience they undergo. But here also there are cases where they approach one another. Let us see if there is any sense at all in which the mystic and thinker can agree in the belief that the mystic experience yields a peculiar access to the divine. Here again we can turn to Professor Dewey for light. He shows us that note of worship and that recognition of a religious object that appears in the thinker when he is sufficiently earnest and profound in his efforts.

> But a mind that has opened itself to experience and that has ripened through its discipline, knows its own littleness and impotencies; it knows that its wishes and acknowledgments are not final measures of the universe whether in knowledge or in conduct, and hence are, in the end, transient. But it also knows that its juvenile assumption of power and achievement is not a dream to be wholly forgotten. It implies a unity with the universe that is to be preserved. The belief, and the effort of thought and struggle which it inspires, are also the doing of the universe, and they in some way, however slight, carry the universe forward. A chastened sense of our importance, apprehension that it is not a yardstick by which to measure the whole, is consistent with the belief that we and our endeavors are significant not only for themselves but in the whole.
>
> Fidelity to the nature to which we belong, as parts however weak, demands that we cherish our desires as ideals till we have converted them into

intelligence, revised them in terms of the ways and means which nature makes possible. When we have used our thought to its utmost and have thrown into the moving unbalanced balance of things our puny strength we know that though the universe slay us still we may trust, for our lot is one with whatever is good in existence. We know that such thought and effort is one condition of the coming into existence of the better. As far as we are concerned it is the only condition, for it alone is in our power. To ask more than this is childish; but to ask less is recreance no less egotistic, involving no less a cutting of ourselves from the universe than does the expectation that it meet and satisfy our every wish. To ask in good faith as much as this from ourselves is to stir into motion every capacity of imagination, and to exact from action every skill and bravery. . . .

The striving of man for objects of imagination is a continuation of natural processes; it is something man has learned from the world in which he occurs, not something which he arbitrarily injects into that world. When he adds perception and ideas to these endeavors, it is not after all he who adds; the addition is again the doing of nature and a further complication of its domain.[5]

This mergence of the individual with the total movement of all things, this sense of dependence upon the whole and participation in the working of this total movement, is surely a religious attitude. There is solemn hope and aspiration and dedicated endeavor and a sense of unity with All, which nev-

---

[5] pp. 420–22.

ertheless is not pantheistic since the unique contribution of the individual is recognized and a certain independent responsibility and power on his part as "one condition of the coming into existence of the better."

Now whatever else God may be, he certainly cannot be separated from the total movement of all things, that total event with which all minor events are continuous. In the mystic experience we yield ourselves up to that event, we merge ourselves with it. To that event, or to the several events that enter into it, all our meanings must refer if they are to have any efficacy at all; and from that event or the several events which enter into it, all our meanings must be derived, according to the philosophy we are now considering.

We become distinct individuals, efficacious in controlling events and contributory to the total outcome of things, only as we develop an operative system of meanings. When we discard these meanings, as in the mystic state, we become merged with events. A system of meanings, and above all a growing system of meanings, is indispensable to individuality and all the values of individuality. We are not for a moment discounting these values. But from time to time we need to discard these meanings, merge with events and so with the Total Event, and thus get a new start. This is what worship does. In its more extreme forms we call it mysticism, but in its rudimentary form it is present in all genuine cases of profound worship. Surely the mystic is justified in the substance of his belief, whatever errors may be involved in his over-beliefs and however mistaken may have been some of his methods.

On the other hand, this surrender to the meaningless flow of events may be cultivated in such a way as to undermine all meanings and cause life to sink to a state of brutishness. If the meaningless state is sought and cultivated as a permanent state and one strives by way of it to escape from all meanings continuously, it is dehumanizing. Such meaningless states, as we have seen, can be had in aesthetic experience or the sexual or gustatory or what not. All these experiences can be cultivated as stages and conditions to the development of a richer and more adequate system of meanings. But if they are cultivated for the sake of destroying all meaning from existence, they are the worst of evils. In other words, the meaningless state has a proper function in human living only as it serves to provide for the refreshing, reconstruction and magnifying of meanings; just as meanings have their proper place only as they serve to control, and give meaning to otherwise meaningless events. The two sides are reciprocal; each is indispensable to the best in the other. Thus work and worship should alternate.

CHAPTER XIII

THE EMERGENCE OF RELIGION

Professor Hocking has said that the chief mark of religion is not utility but fertility. It is the mother of all the great cultural interests of human life. At one time all the arts, using arts in the broad sense to cover the several functions of culture, have at some time or other resided in the body of religion. Politics in its more rudimentary stages was merged with religion and there was a time when men could not distinguish any political life in their society beyond the religious. Education, before it was sufficiently developed to be efficient, and while still blundering and crude, was identified with religion. Science, before it could be called science, before it had perfected its technique, while still vague, confused and ridden with wild illusions, but groping after the truth, was one phase of religion. The same is true of art, of sex love, of agriculture and industrial life, and so on down the list.

This mergence with religion of all branches of culture when at the level of crude immaturity, has brought religion into disrepute. All the vagaries of that pre-scientific groping, with its superstitions, its myths, its blunders, is accredited to religion. In a sense it was religion, but it was religion mothering

353

science. These blunderings, and gropings and superstitions were just as truly rudimentary science as they were religion. They arose because of man's efforts to find the truth. We are not saying that all the myths, illusions, etc., arose out of efforts to seek the truth. We are only saying that that was one motive which has always to some degree, and however sporadically, actuated the doings and dreamings of men. And the working of this motive, in so far as it did work, was rudimentary science or that out of which science developed. But until recent times it was so merged with religion as to be indistinguishable therefrom. This groping has been called science only since it has so defined its field and perfected its methods as to command respect. In other words, the disreputable pre-scientific groping of men after truth has been identified with religion, while the highly respectable and efficient stage of this groping is called science. Such has always been the fate of motherhood—to be identified with the unlovely embryo and the mewling and squalling infant. What has been said of science is true also of art, industry, politics, etc. As soon as these interests became efficient, sufficiently mature to have a technique that would enable them to perform works that could command respect, they became distinguished and separate from religion. Thus religion becomes identified with the crude, the unlovely, the wild guesses and illusions of human beings.

The contempt and doubt that has been cast

upon religion because of this her work of mother-hood, is reflected by William James. He says:

> The cultivator of this science [the history of religion] has to become acquainted with so many groveling and horrible superstitions that presumption easily arises in his mind that any belief that is religious probably is false. In the "prayerful communion of savages with such mumbo-jumbo deities as they acknowledge, it is hard for us to see what genuine spiritual work—even though it were work relative only to their dark savage obligations —can possibly be done.
>
> The consequence is that the conclusions of the science of religions are as likely to be adverse as they are to be favorable to the claim that the essence of religion is true. There is a notion in the air about us that religion is probably only an anachronism, a case of survival, an atavistic relapse into a mode of thought which humanity in its more enlightened examples has outgrown; and this notion our religious anthropologists at present do little to counteract.[1]

The worst of it is, however, that these various branches of culture to which religion gives birth are not delivered completely from the mother's body as an individual organism would be. They are not, of course, individual organisms at all. They break up into various lines of development or non-development. Those lines that do not, that remain crude and archaic, are likely to remain in the body of religion indefinitely. Hence we

---

[1] *Varieties of Religious Experience,* p. 490.

often have a crude science or pre-science allied with religion, which may fight that science which has developed sufficiently to stand as an independent and distinct branch of culture.

But in all this we have the explanation for a very remarkable fact, which is the chief point we want to make. Religion, while being one of the oldest of all the interests of man, is one of the least differentiated and distinct. She has been constantly so merged with her progeny and with the discarded forms of arts which once she reared, that her own characteristic features and form are scarcely known. This is shown by the great difficulty we have in trying to define the precise character of religion. While all other interests, such as politics, industry, education, science, art, and sex love, have been growing out into clear and well-defined functions of culture, each recognized for what it is, each with an "essence" of its own, religion remains still in the dark. It is confused with social service, with morality, with art, with the inertia of tradition, with the illusory play of fancy, with group life as such, with philosophy, with science, and heaven knows what all. It is because religion in actual fact has not differentiated itself from the rudimentary forms of the several arts and sciences, now established or yet to be. She has been so preoccupied in mothering that she has neglected to develop her own unique individuality. Religion is still more or less in that mixed and amorphous condition in which all branches of culture have stood in their rudimentary stages prior to that differentiation and distinction of function which comes with maturity.

It is, of course, well known that the development of human life means this growing distinction of character and function on the part of all cultural interests. Distinction of function does not mean separation, of course, or ought not to mean it and must not, if health is to be preserved. The functional relation means intimate interaction, mutual determination. Two curved lines are functions of one another if the course of one determines the course of the other. The hand and eye are functions of one another if what the eye sees determines what the hand shall touch and what the hand shall touch determines what the eye shall see. This growing differentiation at its best ought to bring about the greatest efficiency and truest freedom and maximum achievement on the part of each interest, while at the same time preserving the closest correlation and interaction of each with all and all with each. This process of functional differentiation on the part of the arts and sciences is one of the marks of progress.

Now the question arises: Will religion undergo this process of differentiation along with the other arts? She has been the slowest of all to stand forth in her unique individuality, for reasons already mentioned. But will she ever stand forth in such manner? Will she always remain, as she has been in the past, suffused throughout the chaotic mass of undefined human interests, or will she unveil a distinct character and function of her own, interacting with all the other interests of man but not confused with them?

Perhaps before this question is answered, a prior question should be put. Does religion have a unique individuality of her own? Is she anything at all save only that mental attitude in which new culture germinates? Is she nothing at all but that confused state of mind and life in which are mixed together, on the one hand the mass of vague, groping, undefined interests which have not yet developed sufficiently to be distinctly recognized and deliberately prosecuted, and on the other hand the cast-off and outgrown forms of arts and sciences which have long since passed beyond the ideas and attitudes which they left behind them in the body of religion? Is religion not a true "mother," but only a mass of eggs, as it were, so that when all the progeny are born there is nothing left but shells and refuse and the spawn that failed to mature? This is the view that some hold, but it is emphatically not our own.

But if religion has never yet merged in distinct character and function, how can we say that she has any? Our answer is that even now religion is in process of so emerging. The thing is going on before our eyes in this age in which we live as it never did before. In the past, religion could serve human life adequately by remaining beneath the surface, moving mightily but indistinguishably allied with other interests; ambiguous, germinating, indistinct. But that time is past and men are turning to religion with the demand that she unveil herself, show her features, reveal her unique identity or else confess that she has none. The world demands such a distinct and characteristic religion to-day because the various arts and sciences have attained such a degree of maturity that they need her, no more as a nursing mother whose identity is merged with their own, but as a guide and companion, with a unique and inspiring individuality of her own, who can stir them to new zeal with her companionship, refresh and enlarge their vision and keep them perpetually young and growing because of the unfolding wealth of experience which religion enables them to discern.

What are the signs that religion is beginning to differentiate itself with a distinctness that it never had before? There are many. Most striking is the intensive study of religion that is being made. Historical, psychological, sociological, and philosophical studies of religion have been prosecuted during the last fifty years to a degree that is unparalleled. The persistent and strenuous efforts to define religion point to the same thing. In other days men were content to take their religion without attempt to define it. But not so now. Religion is becoming highly self-conscious, and self-consciousness is a mark of growing distinctness of individuality. The groping efforts of the church to find its own unique function in society points to the same thing. The separation of church and state, and church and education, the ever clearer line of demarcation that is being drawn between religion and morals, religion and science, religion and art, all tend to emphasize or force to the point of recognition, the distinctive character of religion. Religion as a distinct function is emerging. We know of no greater work to be done in the world today, and no greater need to be met, than just this of bringing religion forth into clarity and distinctness. For only as this distinct and essential function of religion in human life is recognized and provided for, can religion play her rightful part in meeting the deepest needs of men. In the past, as we have seen, it was not necessary to clarify the character of religion in order to enable her to do her proper work. But today it is.

Religion may be used in two senses, to designate its two different phases. Rudolf Eucken has made this distinction and it seems to us it has never been sufficiently recognized. He distinguishes the two by calling one *Die universale Religion* and the other *Die charakteristische Religion*. In one phase, religion is the mass of germinating culture consisting of embryonic arts in all stages of development mixed with the cast-off, outgrown forms of arts that have long since outgrown the body of religion.[2] In the other phase religion is a characteristic function distinct from the other functions. This does not mean, of course, that religion in this latter sense does not pervade the whole of life, as religion always should. We have already explained the use of the word function sufficiently to

---

2 We here use the word culture in the broad sense to designate all that portion of the life of man that distinguishes him from the beast, in a word his spiritual life.

show that we do not mean by it something separate from the rest of life.

The importance of this distinction will not appear until we consider some of the studies that have been made of religion. We have already quoted from James to show the dubious and contemptuous attitude that so easily arises toward religion when its history and anthropology is studied. What these studies put before us is not characteristic religion, not religion in its unique character and distinct function, but it is simply germinating culture. In a certain sense it can be said that what they study is religion; but it is equally true to say that what they study is rudimentary culture, embryonic science, embryonic art, and politics. What one should condemn in these early stages of life is not religion as such, but the cultural interests of man when at these rudimentary stages of development. It will help to free religion from this unjust treatment, and clarify the unique character of religion, if we glance at some of these historical studies. There is perhaps nothing better in this field than George Foote Moore's *The Birth and Growth of Religion*. We shall take it as the best example we can find of such a study.

He does not try to define religion but he does find it necessary to set up certain marks by which to recognize it when he finds it. These marks are four:

(1) Man's belief in certain powers that do things to him.

(2) Belief that these powers are actuated by motives similar to his own.

(3) Belief that he can induce these powers to behave in such a way as to help him or refrain from hurting him.

(4) Action according to these beliefs.

By powers are meant simply "things that do things" without any attempt to define their nature. Their nature will be defined variously in different religions. What makes these beliefs religious, however, according to Mr. Moore, is not their character as beliefs, but the way they are acted upon. One must stake his welfare on his belief and act accordingly. Otherwise the belief is a philosophy, a fancy, a myth, or a science, but not a religion.

His welfare that he commits to his religious beliefs, the character of the beliefs themselves, and the way he acts upon his beliefs, are all determined by his wants. If he wants protection from wild beasts, from storms or cold or parching heat, he will construe the gods in such a way as to make them fit and able to provide such protection on condition that he himself acts in the proper way toward them. If he wants his herds to multiply, his god may be a bull. If he wants crops to grow and yield, his god may be a rain god or a river god.

The impulse giving rise to religion, says Mr. Moore, is that of self-preservation, which includes preservation of the family, the tribe, the species, the land or whatever concerns and constitutes that system of interests which makes up the self. At first this impulse is chiefly concerned with the attainment of material goods and avoidance of material ills. But in the course of time it is "self-realization," the development of all the powers of the self, the attainment of a transcendent self. This self-realization becomes the governing interest and for it biological existence may be sacrificed. Without it all things else have no value. When this interest becomes paramount the God (or gods) become some sort of transcendent self (or selves), an over-self, through whom this self-realization can be attained.

The first step in the development of religion, traced by Mr. Moore, is the transition from dealing with things as powers in themselves to dealing with the soul or spirit back of these things. At first the back-switching branch, that lashes you in the face, or the lightning that frightens you, or the wandering stream which led you to your quarry, was itself a malevolent or benevolent thing with which you must deal directly. The branch itself, or the lightning or stream, you must control or pacify, or otherwise act to win its favor or escape its malice. But in the course of time it is not the branch itself or stream or lightning, but the spirit in the branch with which you must deal.

This belief in souls arises from two sources; observation of the dying and dreams.

The man who was energetically doing things, now lies cold and still. How came this chnge? It came when he ceased to breathe. It must be, then, the breath which has passed out of him which wrought the change. His breath, then, was that which animated him, which did things in him. His self, that which does things, is the breath. Or again, it is when the blood flows out of him that he ceases to do things. It must be either the breath or the blood, or both, or sometimes one and sometimes the other which is the doer of things. Thus soul or spirit is the word by which he designates the doer in the body, and is likely to be conceived as another more ethereal body made of breath. So it comes about that when the branch or stream does things which attract his attention he thinks it must be some spirit in the branch or stream which does it.

Dreams further augment this process of filling the world with spirits akin to the "soul" of man, and making spirits rather than ordinary sensible things the chief objects of human concern. In his dreams he deals with men and animals and other objects. In dreams he wanders in regions remote from the place of his sleeping body. How is this possible unless he has a double, a soul, which can leave the body, wander to these distant parts, and interact with other objects. Also in dreams one sees men and beasts which come to where one is sleeping, which may converse and even struggle with one; and yet on awaking one may discover that these men and beasts in bodily form have never been present. Hence they also must have souls that leave the body and wander abroad.

Thus there gradually develops the notion of spirits which animate objects whenever these objects do anything to attract especial attention. Hence it is upon these spirits that one's welfare is dependent. If one would get food or escape harm, ward off danger or be grateful for good received, it is to these spirits that one must turn. It is they that support one on the floating ice, or bring the fish, or cause the log to span the stream. Whatever is done they do, for they are the doers. But that does not mean that early man generalizes and asserts that all the objects in the universe embody spirits and all the happenings in the universe are done by spirits. He does not ordinarily generalize at all. He does not think about the universe as a whole. He thinks only of this particular happening and that. These, on those particular occasions when they most conspicuously affect his interests, he believes to be animated by spirits. But what the universe as a whole may be, may never enter his head.

Before pursuing the sketch further there are two points of criticism which we must raise. The first applies to the concept of religion here implied. On what grounds can one call all these beliefs, inferences, illusions and practices religious? Mr. Moore says that the belief and practice can be called religious if the individual commits his welfare to the belief and acts accordingly. But surely we have scientific beliefs to which we commit our welfare and act accordingly. We have ordinary, practical, everyday beliefs on which we constantly stake our welfare, and yet we do not recognize anything distinctively religious in them. No, we

must insist that these developing beliefs which Mr.
Moore has so accurately sketched are no more rep-
resentative of rudimentary religion than they are
stages of rudimentary science, rudimentary art,
rudimentary social life, rudimentary love, and
rudimentary culture in general. The process by
which men come to belief in spirits is a scientific
and artistic process every bit as much as it is a religious process. Of course if one insists that these
products of imagination cannot be called scientific
and artistic because the latter necessitate a special
technique and distinctive function, then we can
say the same with regard to religion. One can insist that these processes are not religious because
they do not reveal that special function and technique which we find in the religion of today. The
blade cuts both ways. Either this process of developing beliefs is rudimentary culture in that stage
prior to the differentiation and technical proficiency
of the several arts and hence is just as much embryonic science, art, and social life, as it is embryonic
religion, or else it is none of these, neither religion
nor science nor art. We admit that in all this there
is certainly an element of religion just as there is an
element of science and art and social life, etc. But
we cannot identify it all with religion any more
than with the others.

The second point of criticism is with regard to
the motive or impulse that gives rise to all these beliefs and practices. Mr. Moore says it is the impulse
of self-preservation. We feel that that is a very inaccurate statement. The lower animals are actu-

ated by the impulse of self-preservation, but they do not develop any such world view, or religion, if one wishes to call it religion. They experience the death of their associates and the higher animals have dreams. Why do they not develop beliefs in spirits and try to pacify, persuade, or control them? Plainly the impulse to self-preservation is no explanation at all.

Why, then, does man go through this process of developing a rudimentary culture, one ingredient of which is religion? We cannot ask, Why does he develop a religion, for reasons already stated. He does not develop a religion as something distinct from rudimentary, germinating culture, in which there is mixed together as much science, art, social life, and love as there is religion. But why does he do it? The answer to that question would lead us very deep into the nature of man and into the understanding of religion. Santayana would say that it is due to the lyric quality of human consciousness which spins out fancies for delight. John Dewey would say it is the exuberant reminiscencing to which man is so prone. But we do not believe these answers go to the root of the matter.

The child, the savage and the man in general develops a boundless efflorescence of beliefs, fancies, myths, and speculations because he is organically so constructed as to be responsive to a far greater number of stimuli than those which control his adaptation to immediate environment. Putting the same thing in other words, it is because man is endowed with a far greater number and diversity of impulses than can be fulfilled in ordinary adaptive behavior. It is these suppressed or inadequately expressed impulses, responsive to the rich fullness of the world about him, that give rise to all these groping beliefs concerning this vague, dim, rich fullness of the world. Still another way of expressing the same thing is to say that man is aware, prior to sophistication, of far greater wealth of immediate experience than he is able correctly to interpret or use in adapting himself to his recognized environment. Because of these surplus stimuli, surplus impulses, and surplus wealth of experience, he is given to wondering, speculating, imagining, and groping out into the unknown, striving to solve the mystery of that total impact of the world upon him which so far exceeds those few features which serve to guide him in meeting the routine requirements of life.

Now it is our claim that religion is precisely our response to the undefined significance of this total wealth of experience when we take it as a single datum signifying the supreme and total object with which we have to do in all the conduct of our lives. This is religion, we say, when religion becomes sufficiently differentiated from the several branches of culture to show its unique character and distinctive function. The several branches of culture, on the other hand, are distinguished from religion in that they select from this wealth of surplus experience certain features or data which they circumscribe as their own, and with which they deal by a special technique which is also their own. The several sciences do this. The several fine arts do this. Romantic love does this. Friendship does it. So also politics, etc., etc. Underneath, or running through, this wealth and surplusage of experience which provides the materials for the several branches of culture and for religion, there are those common everyday features which guide us in our biological adaptation to the immediate environment. It is because the lower animals scarcely attend to anything more than these common everyday features, that they do not develop a culture or a religion.

We have just defined religion and the branches of culture as they are when they become differentiated and distinct in character and function. But in the early stages of life as portrayed by Mr. Moore, and even to a large degree in modern life, they are not differentiated. Differentiation is a matter of degree, and in primitive life it is at the minimum. Consequently at that time religion was not what we have just described it to be nor were the branches of culture. They were mixed together. That means that early man did not respond to the wealth and surplus of experience as a total datum signifying the supreme and total object with which he has to deal. Nor did he select from out this total wealth certain features which were allocated to certain special techniques. He selected certain features but not according to any special plan. He reacted to the surplusage of experience in a more or less merged and inchoate form, and yet not as a total datum signifying one supreme object. His position was a sort of compromise between the religious attitude on the one hand and the attitude of the several arts and sciences on the other.

There is a still third criticism we would make of Mr. Moore's treatment, which supports the view we have just set forth. We believe Mr. Moore makes primitive man altogether too practical. He represents him as thinking, doing, and believing, always with a view to satisfying certain practical needs such as food and shelter. Both Santayana and John Dewey have shown this to be a fallacy.[3] Because we are living in an age and in a country dominated by practical interests, especially economic, we are prone to think that all men of every age were so controlled. But children and primitive men are not preëminently practical. They are not chiefly concerned with matters of food and shelter. Early man by nature is a dreamer. He only gradully learns to work and think with a view to satisfying his material needs. He is not nearly so diligent in the pursuit of material goods as are the lower animals or as are many men in civilized conditions. He must learn to work, to constrain his imagination, to engage in directed thinking. And he can learn this only as he has painfully built up a system of tradition that holds him to such endeavors. The savage does not have such a system of tradition ordinarily. To be sure he is often hard pressed. But he would rather play and dream and die than work, or bend his thought to practical matters. He is not nearly so much of an economic man as Mr. Moore and others would represent him to be.

---

[3] Dewey's most effective treatment of the matter is in his *Reconstruction of Philosophy*.

Of course early man's imaginings are shaped by his wants. But what are his wants? Food and shelter, yes. But these are only two among innumerable others. And among his many wants, strongest of all in the case of some, is the want of venturing forth into the dim unknown that encompasses him. This he can do most readily and safely by exercising his imagination. So he commences to develop beliefs. These beliefs are religious in so far as they pertain to the significance of the merged and total datum of experience prior to analysis and discrimination of distinct factors within it. For these several factors do not signify God, but only the total indiscriminated event can involve God. Some of his beliefs may approximate the status of being beliefs concerning this total event, for he has not learned to analyze and discriminate within it those factors which signify separate objects. But neither has he learned to apprehend the total event as the single unanalyzed passage of nature. Hence he is neither thoroughly religious nor thoroughly scientific, and he is scarcely any more the one than the other. He neither discriminates the data accurately nor merges them completely into a total datum. He does not carefully investigate the significance of distinct data nor consider the significance of this total datum. Rather his mental attitude wavers back and forth between these two extremes. Hence he is neither scientific nor religious but a mixture of the two which may ultimately develop into both, but for the time is neither. It is apparent, however, that these two lines of development must move in opposite directions.

Religion is just as "instinctive" as science and art, but no more so. The only reason the view has been held that religion is more "instinctive" than politics and cultured love, is because religion has not been defined in thought as clearly as these latter. All the crude inchoate stage of culture has been lumped together as religion. Of course when that is done religion appears to be more primitive and closer to original human nature than the several arts and sciences. We have tried to show how this misunderstanding so easily arises.

As a matter of fact one must learn to worship, to be thoroughly and consistently religious, just as much as he must learn to love. The sex impulse does not need to be learned, but love does. Gregariousness does not need to be learned, but political life does. An imaginative reaction to a back-slashing branch does not require a very great amount of learning, but to enter into the distinctive religious experience does. To be thoroughly and consistently religious requires a gradually and painfully acquired system of proper traditions, and the same is true for thoroughness and consistency in scientific method or artistic production.

Religion as something unique and distinct, we have said, is our response to some significance of the total datum of experience, or, what Whitehead calls the total undiscriminated event which may enter awareness under favorable conditions.[4] But to be religious in this way requires two things which

---

[4] Whitehead, A. N., *Concept of Nature.*

are interdependent. One must somehow become aware of this total event and, secondly, he must find some significance in it. He can scarcely become aware of it unless he does apprehend it as significant, however undefined the significance may be. To find significance in this total event means to have certain beliefs concerning it. Such beliefs are truly religious. But they can arise, it is quite plain, only as the result of long development. They can appear only in "the fullness of time." A fully developed and differentiated religion may be said to consist of two parts: (1) the technique, ceremonies, etc., by which one enters into an awareness of this total event in which we "live and move and have our being"; and (2) beliefs concerning this total event. It is plain that the primitive man cannot have religion in this sense, but he can have the beginnings of such a religion, just as he can have the beginnings of science and political life.

To find religion in its clearest and best defined form, we must go to its highest development, rather than its lowest. Only there does it begin to become differentiated from foreign admixtures. Nothing better has been said upon this, we believe, than the words of Rudolf Eucken, which we must quote:

Wer den Wahrheitsgehalt der Religion Ergrunden möchte, der braucht weder ihre verschwinden zeitlichen Anfange aufzuspuren, noch ihr langsames Aufsteigen zu verfolgen, er darf sich so fort auf ihre Höhe versetzen. Denn erst hier erlangt das Wahrheitsproblem volle Klarheit and zugleich eine zwingende Kraft. So Kummert uns

nicht das Zauberwesen, das die anfangsstufe der
Religion begleitet and beherrscht, so braucht uns
auch nicht die Religion als ein bloss Stuck enier
Volkskultur and als Naturmmthologie zu be-
schäftigen.   Sondern unser Problem beginnt erst
da, wo sie eine eigene Welt erzeugt, diese allem
ubrigen Leben entgegenhalt and es von ihr aus
umgestalten will, umgestalten dadurch, dass sie
den Menschen inmitten seines eigeneen Kreises
eine unsichtbare Ordnung, ein ewiges Sein, ein
ubernaturaliches Leben vorhalt, and dafur seine
Seele erlangt.[5]

Let us follow further the excellent sketch of Mr.
Moore's, not with the hope of finding the distinct
form of religion in these historic stages but in order
to show that religion is throughout in process of
emergence, that it is indistinguishably mixed with
other interests, moving mightily beneath the surface of life, hidden by the flux of many things.

Magic and religion cannot be set over against one
another, says Mr. Moore, for they merge. However, religion in a higher and purer form may be
contrasted with magic. Magic is that practice which
is designed to coerce the spirits, while this higher
stage of religion persuades and promotes fellowship
with the spirits.   But he adds: "Many rites even in
the higher religions are survivals—sometimes symbolized—of magical performances; and what is of
greater moment, acts which in their origin were, by
our definition, purely religious, often come to be
regarded as so infallibly effective that the element

---

[5] *Der Wahrheitsgehalt der Religion*, p. 1.

of persuasion recedes and the rite duly performed is believed, at least by the vulgar, of itself to accomplish the desired result—that is to say it lapses into what we have defined as magic." We might add that we have known Christians to use "in Christ's name" or a similar phrase as somehow coercive on God and hence a form of magic. And it is quite a widespread notion among people who are not professedly religious, as well as those that are, that conformity to the ten commandments, or obeying the Bible in some other particular, will promote prosperity by coercing fate or providence. All of which goes to show that religion as it shows itself today is not clearly differentiated from an undeveloped science, an undeveloped morality, and other undeveloped forms of culture.

The next step in the development of religion, which Moore calls the emergence of Gods in the proper and fully developed sense, occurs when civilization arises with the pastoral and agriculture life, with increasing density of population and complex social organization and with permanent responsible rulers exercising many and great powers over the lives of common men. With the domestication of animals and drawing of livelihood from them, men become aware of dependence upon certain outstanding features of nature such as seasonable rainfall associated with sun, sky and stars. Thus the innumerable, capricious and unorganized spirits that thronged in the earlier stage, become more or less concentrated or subordinated to the spirits embodied in these outstanding features of nature. Agri-

culture still more concentrates attention and the
sense of dependence upon rather well-defined forces
of nature. The shepherd can drive his flocks elsewhere, but the agriculturalist must continue dependent upon the same river, the same winds, and the
same climatic conditions. In regions where agriculture first arose nature was bountiful and while
disappointments came often enough to keep alive
his sense of dependence, he came to think of the
powers as friendly to him and his community.
Thus he put his trust more in the worship that persuades, or wins and holds the favor of, a friendly
god and less in the magic that constrains unfriendly
or indifferent powers to serve him.

The next great step in the growth of culture (we
can scarcely call it, as Moore does, the growth of
religion, except as it is merged with the cultural
process) is the idea of God not merely as protector
and benefactor but as ruler and lawgiver. As property owner and member of a complex social order
man's welfare became more and more dependent
upon law and order and his "religious beliefs" were
shaped accordingly. The rise of great earthly
rulers gave shape and body to the idea of God as
supreme lawgiver and law enforcer. Thus religion
became more and more a matter of morals and of
law; God became humanized, civilized and moral.
The development of a cult of regular and orderly
worship with a priestcraft, an established ceremony
and temples or other sacred places, further aug-
mented this process of making God the ordainer and
sustainer of the laws and practices of organized

society and religion the great sanction of morality.

The development of appropriate myths carried the process still farther. "These stories about the gods help men to imagine them as magnified and glorified men and women with individualities of their own that are not simply the reflections of their spheres of activity. Even the less edifying tales only make the likeness stronger and create human fellow feeling. In the progress of civilization such myths become repugnant to their more refined taste or a more elevated morality; and objection is made to the scandalous chronicles of the poets. . . . But here again the very idea that the gods should set a good example to men is the last consequence of their complete humanity."[6]

But when we speak of religion as developing from one stage to another, we do not mean that the later drives out the beliefs and practices of the earlier. The earlier persists either alongside the later or in some sort of amalgamation with it. Worship of praise and petition does not drive out magic. The morality of God does not drive out belief that the gods still practice that which would be immoral in any living being. The supremacy of one god does not drive other gods beyond the pale of human concern.

But when some god or gods became identified with the moral requirements of the group, transgression of these requirements became sin, *i.e.* personal offence to God. This is a matter of great importance to the advance and support of civiliza-

---

[6] *loc. cit.* p. 56.

tion. The small, isolated, homogeneous group had but one harmonized system of mores which moulded the conduct and belief of the individual with almost irresistible force. But:

> With the growth of cities and nations of amalgamated populations and complex interests, the compulsive force of community opinion relaxed. Advance in knowledge of the world and the workings of nature undermined many primitive beliefs. By taking morals into its sphere and making them part of a divine law religion furnished the only possible substitute for the sanction of the primitive mores; a substitute effective as long as the authority of religion itself was not challenged. But on the other hand religion with its characteristic conservatism, gave a degree of fixity to morals as well as to the rituals of the past, the unsorted accumulation of ages. What we regard as moral was inextricably entangled with nonmoral interdiction. . . . Religion was not made moral, but morality religious; and religion thus often interposed a formidable obstacle to moral progress. This is peculiarly the case where law is fixed in sacred scriptures containing a closed canon of revelation, which admits no addition or subtraction, no change, and thus gives the stamp of finality to the institutions and laws and moral standards of a past which religion is forbidden to outgrow.[7]

Moore goes on to sketch in a masterly fashion the development of these beliefs and practices through further stages of civilization, until he reaches the

[7] *loc. cit.* p. 76.

great soteric religions, Zoroastrianism, Judaism, and Mohammedanism, which are concerned with the conquest of evil in nature and society, both in this life and in a world to come. Then he considers those other soteric religions, the mysteries of the Hellenistic-Roman world, and certain religions and philosophies of India and Greece which are concerned to deliver the isolated individual from the evils of embodied existence in a world of matter and sense.

Finally he takes up Christianity, showing it to be a unique synthesis of all the higher religious, moral, and philosophical aspirations and endeavors of the Mediterranean area for long centuries—"the religious legacy of the ancient world to the ages that were to come." The original and germinating nucleus about which this synthesis of Christianity was organized was devotion to the personality of Jesus as expressed in his teachings and doings. This synthesis is a cord of three strands: Jewish ethical monotheism; Hellenistic soteriology, profoundly modified by the Jewish element; and Greek philosophy, which not only constituted the formal principle of Christian theology, but made large contributions to the material element. "The intellectual victory of Christianity over all the rival salvations of the time was due to the fact that it alone offered not merely a way of salvation but a philosophy of salvation."[8]

We believe this survey of the birth and growth of religion shows most plainly the confusion be-

8 *loc. cit.* p. 178.

tween religion and general culture when the latter is in its early stages of development. It is truly a history of religion if we take religion in that broad general sense which makes it comprehensive of all germinating culture. But it does not portray religion as a unique and characteristic function in human life. For instance, when Mr. Moore singles out Thales as the man who broke with the past, cast off mythology and adopted the scientific method of procedure, he is perfectly right. But in so doing he is defining science in such a way as to exclude its origins. All those beliefs and practices which he so finely traces before the time of Thales, and calls religion, are what finally eventuate in the science of the Greek. On what grounds does he say that science suddenly emerges at this point instead of gradually developing through all that efflorescence of the human mind which went before? He does so on good grounds. He does so because science is so defined as to exclude all those earlier processes. But why not define religion in such a way as to exclude all these crude and vague beginnings also? Because religion has not, as science has done, emerged from this chaotic, germinating mass of spiritual endeavor to such a degree that her distinctive lineaments can be recognized by all. But that does not mean that she has no such lineaments. No doubt she stood forth many times, clearly and distinctly, even in those days of early savagery, just as scientific method may have appeared sporadically many times before the days of Thales. But only with him did it become an established institution that could be regularly perpetuated throughout successive generations. Religion, pure and distinct, has appeared many times, no doubt, in individual lives. But it has not been clearly and universally recognized for what it is, deliberately cultivated in this pure and definite form, and so perpetuated throughout successive generations by means of established institutions. That which has been recognized as religion, deliberately fostered and perpetuated as such, has been a confused mass of outworn husks and germinating buds of spiritual life. Religion pure and simple, with its own function in life clearly distinguished from all others, has not yet stood forth in our midst. But we believe its appearance is near at hand. We believe its steps are even now at the door and its hand on the latch.

Religion is man's endeavor to find that adjustment to God which will yield most abundant life. For God is precisely that object, whatsoever its nature may be, which will yield maximum security and abundance to all human living, when right adjustment is made. With this definition of the term it cannot be doubted that God exists. The exact nature of God is still problematical and may be for many years to come. The most important problem that can engage the mind of the human race is that of discovering what God fully and certainly is. Here again our definition of the term makes this statement self-evident. We believe much valuable work has been done toward the solution of this supreme problem of all human living. But still more valuable work awaits the doing.

Some may think this most important object of human concern is society. But a little thought will show that it cannot be merely society in general which is of supreme importance. It can only be some particular instance of society, one's own social group, for example, and that "kingdom of God" into which it may be developed. The object of supreme concern, from the standpoint of society, must be those present and actual conditions, adjustment to which will bring to pass the most desirable society and sustain it in so far as it may be attained at any given time. These present and actual conditions which are most critically and ultimately important, taken collectively, constitute God. That is what our definition of the term implies.

Some may think this most important object in the universe is a certain ensemble of chemical elements. Still others may think it (or He) is electricity or some other such pervasive form of energy. The moon-mad lover may react to his lady as holding this place of highest value; and the drug addict may treat his opium as having the status of God. The prophet may think God is a superhuman mind that is somehow operative in the universe. Some may think God is the universe taken in its organic totality. We do not think any of these statements concerning God are satisfactory. But perhaps all of them afford some glimpse of the truth and God can be known for what he is only when these glimpses have been worked out into a comprehensive vision. But God, we know, is the most important object that can engage the attention of man.

The only hope of bringing human life to its largest fulfillment depends upon finding God and the right adjustment to him. The very meaning of the term God, as we use it, shows this to be the case. The fact that human life continues and grows shows that in some measure we have found him. But we must find him more completely. And, what is much more important, we must find how to make that adjustment to him which will bring human life to its maximum security and abundance.

The greatest creative enthusiasm of which human life is capable can be awakened only when a man gives his whole attention to the object which is, in fact, the condition of greatest enrichment to all human living. Only as a man exposes himself to the full stimulus of this object can his utmost capacity for response be awakened. This attitude toward the most important object in the universe is worship. And worship is the heart of religion. From it arises the largest creative endeavor humanly attainable.

No matter what a man treats as the most important thing in the universe, he is worshipping if he puts himself in that attitude toward it in which he can experience the maximum stimulus it can yield. He may be worshipping an idol, but it is worship. If it be the true God, however, or that which approximates the true God, his worship will arouse and organize his impulses for the farthest swing of constructive achievement of which he is capable. Such, we believe, is the rightful work of religion in human living. Its function is creative.

INDEX

385